Experience

Experience

Donald J. DeGracia, Ph.D.

PlaneTalk
About this and the other worlds

DeGracia, Donald J.
Experience
ISBN 978-1-312-55137-4
final version

email: dondeg@compuserve.com
        ddegraci@med.wayne.edu

Visit me on the web at:
www.dondeg.com
dondeg.wordpress.com

Ordering Information:
Quantity sales, special discounts on quantity purchases by corporations, associations, and others are simply not available. This is a print on demand book. I do not control the economics of any of this. There are no details, therefore no reason to contact the publisher at the address above. Orders by U.S. trade bookstores and wholesalers: Please do not bother to contact the email addresses above. Lulu has bulk discounts listed on their web site. Instead of pens and coffee mugs with your company's logo, buy many copies of this book to hand out at trade shows and expos. Make an impression!

Printed in the United States of America. With Freedom and Justice for All.

This work is dedicated to the unceasing prattle, noise, and hot air of all the philosophers and thinkers whose influences are now subconscious and ubiquitous in my thinking.

Contents

Introduction

This book is a collection of blog posts from summer/autumn 2014, collectively called *Experience*. The work is intended to help people who have been encultured in the Western way of thinking to understand why anyone would wish to practice yoga. Given the glut of yoga-related materials on the market, what is the need for such a book? It is because the West still has not fully grasped the essence of *real* yoga.

The framework of yoga is Hindu philosophy. Expositions of Hindu thought in the West have been available for at least 175 years. Initially these were confined to small groups, mainly indologists and esoteric groups such as Theosophists. But in the intervening years, this knowledge has diffused throughout Western cultures in increasing measure, albeit slowly due to the prejudices of Western thought.

A notable turning point was publication in 1901 of Richard Maurice Bucke's *Cosmic Consciousness*, a compendium of historical examples of mystical experiences. By 1945, Hindu concepts had made it to the avant garde intellectual layers with publication of Aldous Huxley's *The Perennial Philosophy*. This line of development intellectualized Hinduism in a characteristically Western fashion, which is to say, in a comparative way against other forms of mystical thought, both Eastern and Western.

A second line of development manifested when Swami Vivekananda spoke at the Parliament of the World's Religions in Chicago, IL in 1893. This event stimulated a significant interest in the West of the ancient practice of yoga.

A third important line of development was the discovery of LSD, lysergic acid diethylamide ("acid"), by the Swiss chemist Albert Hofmann at Sandoz in 1938. This led eventually to the "psychedelic 60s" in the Western countries, where people widely experimented with the mind-altering properties of this and similar psychedelic drugs.

Together, these three lines of 20th century development: (1) the diffusion of Hindu philosophy to Western intellectuals, (2) the introduction of yoga to the West, and (3) the widespread experience of altered states of consciousness made possible by psychedelic drugs, form the historical basis of the present work.

However, things never follow a straight-line path of evolution. All three lines were eventually derailed. Interest in Eastern philosophy waned as the computer revolution took off. Kids today are more interested in the latest iPhone than in the *Bhagavad Gita*. In the West, yoga devolved into a mere form of exercise and is now associated with sports medicine. Psychedelics were made illegal.

Meanwhile, Western societies have significantly deteriorated. Deindustrialization of first-world countries and the rise of third world economies have generated serious economic turbulence. First-world countries now run on <u>fictitious capital</u> instead of tangible wealth. Constant unemployment, ever-growing inflation, and displacement of white-collar jobs have eroded living standards. The present political climate can only be characterized as decadent and insane, where the power elite ever more openly display their brutality and depravity.

The legitimized institutions of Western civilization – governments, law, medicine, intellectual property, the mass media, the universities and traditional academic learning – are in a state of flux not seen since the beginning of the industrial revolution.

A key factor in all of this is the rise of the internet. When it went public in the 1990s, the internet was envisioned as a source for a "new economy". Instead, the internet has emerged as a threat to the status quo in ways no past political revolution could ever imagine. Many of the particulars are well-known and I won't repeat them here.

However, an important general effect of the internet is that the power elite have lost control of the "great unwashed masses" via the centralized mass-media propaganda machines of the 20th century: TV, newspapers, and movies. The internet has breathed new life into the old idea of "marketplace of ideas". The internet now reflects the natural cacophony of human social and mental life.

The plasticity and malleability of what is considered "truth" has become evident for all to see. However, truth has always been malleable. But in calmer times, when jobs are aplenty and social institutions, in the aggregate, work for the social good, people are less willing to hear that life is constant change. When society becomes chaotic, people's minds become open to the message of change and flux.

But beyond the change, there is something that doesn't change.

The Western mind has always struggled to grasp the unchanging essence behind changing appearances.

The Eastern mind relaxes and thereby reflects this truth.

The reason the Eastern mind can reflect the eternal is because it sees through the mirages of the transient. Meanwhile, Western civilization is obsessed and preoccupied by the transient patterns.

This contrast between the transient and permanent, with emphasis on the transient, is the theme of *Experience*. What sets *Experience* apart from similar studies is that, right from Chapter 1 we take seriously the consequences of altered states of consciousness. The fact that these exist and we can contrast them to our so-called "normal" state of consciousness is a key pillar of the logic used throughout.

Most expositions on the meaning of life ignore the fact that humans are not confined to only one domain of awareness. Such views are therefore inadequate and incomplete, so much so that they lead to wrong conclusions about the human condition.

Yoga does not hide or shy away from the fact of altered states, but embraces and harnesses them for a much deeper purpose.

But altered states are not a panacea. As will be discussed, they are a deeper form of illusion; an amplification and magnification of the illusions of surface consciousness. The illusory nature of altered states is known to yoga, which starkly warns of their beguiling influence.

Why yoga? Because it holds out the promise to actually *experience*, as opposed to merely intellectually know, the answers to the perennial questions of Western religion and philosophy. The fact that the answers to such questions can be *experienced* is a shocking revelation to the typical Western mind and is met with skepticism and denial.

Therefore, *Experience* seeks to explain to people with a typical Western outlook why the yogic approach is not only understandable, but inevitable. As we show, yoga is *not* body exercise, not by a long shot.

A summary from later in the book is repeated here to provide an overview of the flow of the book:

- **Chapter 1**: Words and ideas cannot capture experience but they can, more or less perfectly, reflect experience.
- **Chapter 2**: Experience consists of always striving, but never quite getting there.
- **Chapter 3**: This always-striving seems to be present in all natural system we can observe.
- **Chapter 4**: Since experience consists in always-striving, images in our mind of completeness and fulfillment are just mirages.
- **Chapter 5**: Even altered states of consciousness offer no solace. They do not improve the condition of always-striving found in the surface mind but only offer more of the same.
- **Chapter 6**: Always-striving reflects the incomplete nature of potential infinity. Manifest experience teases us with promises of completeness, but never delivers.
- **Chapter 7**: The always-striving of experience is due to us projecting our being onto our perceptions. This affects our philosophical understanding and practical everyday experience.
- **Chapter 8**: Yoga is based on two facts: (1) we are conscious, and (2) we transition between waking, dreaming and non-dreaming states. Success with yoga reveals that "normal" experience only seems to be, and that behind this seeming there is a true reality that simply is.

Thus, to use the old cliché, *Experience* brings good news and bad news. The bad news is that the ways of Western civilization are a preoccupation with mirages and illusions. The good news is that there is an alternative.

Hopfully I am successful conveying the message. If not, well, I tried.

Don DeGracia, 2014

Experience

1. Experience

This one has been kicking around in my mind for a long time: What is experience?

Somewhere and somewhen about 1987, I did five hits of acid (yes, I was in college when these things are done, and yes, it is part of studying altered states, so just grow up). The whole trip peaked with me completely focused on the question: "what am I?"

I cannot even begin to express in words the sheer intensity of focus, desire, anxiety, the sheer raw intensity upon which my mind was focused on this question. My mind raced. It raced and raced, and spun and spun and went round and round, faster and faster, through picture after insight after concept after profound realization of all the definitions and ideas I had learned, and some I realized right then, about what being human means.

My mind spun and spun and spun until it ripped open. Or burst, or exploded; in some literal sense. It's hard to say exactly what happened, seeing as there is no worthwhile concept or definition in Western thinking of what the mind really is. And I say this being, even at that time, a student of occultism, mysticism, science, philosophy and just plain intellectual stuff.

The intense wonderment and amazement at the very mystery of my own existence culminated in a stark insight that today is one of the foundations of my outlook: it is impossible to know what I am because knowing is something that happens in the mind, and, my mind is only a subset of whatever I am. Whatever I am is greater than what my mind is because my mind is inside of me.

That is, any idea I had about what I am was just that: an idea. As such, it was just some thought inside of me. It was just an echo, a reflection. Just an echo. Just an echo of experience, inside my experience. I didn't even know what "me" meant anymore, or what "I" means, or "you". I still don't either.

All of a sudden, I understood fully why Zen masters just point at stuff and don't say anything.

In the culmination of all this, I had a really bizarre insight, a vision if you will, of all the philosophers up and down history: Socrates, Plato, Kant, Leibniz (yes, my hero Leibniz), Swedenborg, Blavatsky, Leadbeater, Krishnamurti, Alan Watts, Carl Jung, and so many others whose ideas had influenced me. I saw in a flash what a total bunch of fucking idiots they were for thinking that we could capture in words and ideas what we are. I saw how all their words, their unceasing prattle, were just a bunch of noise and hot air being blown around.

The observant Reader hopefully is catching a very deep irony here.

And so I realized that there is no definition of what we are. Period. None. Thinking that we can capture what we are as words and ideas is like trying to pour the ocean into a cup. Sorry, there is no cup big enough. Same with the mind. It's just too small, too limited in how it works to somehow encompass our...what is a good word?...our experience.

That is what I realized back then. There is only experience. Inside of our experience, as a part of our experience, one of the things we experience is called mind. It's not the only thing we experience. I, like everyone else, experience this whole complex, truly undefinable thing we call "our life". And no words, no ideas, are capable of capturing my very experience.

As if somehow I can string together words and ideas and literally re-create my very being and experience. It is an absurdity, like the idea of one hand clapping.

Another way to say this is that ideas are always inside of the mind. They do not leave the mind. I am sitting here right now looking at my computer monitor, looking out my window at the trees. Those things exist independent on any idea in my mind. As such, I have no idea what they really are, only what I think they are. My ideas are not "out there", outside of my mind; they are not sitting there next to the computer or the trees.

So, when we have some idea about what we are, or what the meaning of life is, or whatever, it is just some picture or set of words that elicit a meaning in the mind. The mind is always inside of itself. It is like a dictionary where words are defined only in terms of other words. A closed system. It is like a hamster running on one of those wheels. It goes very fast and it goes nowhere. That is the nature of thinking in general. That night I saw myself going round and round and round in my mind. And that was all I was doing.

This is the nature of thinking: we move very fast, round in round in circles, and we go nowhere. Thoughts do not allow one to leave the mind.

So, from that moment on, experience became the main catch word for me. I didn't know what it meant (and still don't). But the word "experience" became like a sign post, a symbol that points to my life itself.

With this grand insight, everything changed in my experience. I saw how, before my explosion, I had the delusion that my ideas were somehow superior to my experience. That the words and ideas I spouted were somehow, in some sense, better, more real, than my actual life. Now I know it doesn't matter what I think or believe: tomorrow I could die, and I still have to eat, sleep, and shit every day.

The fact that I called this way of thinking a "delusion" clearly indicates the change I underwent. I lost the expectation that everything can be defined, that everything can be understood. In fact, I came to the realization that, really, nothing can be defined. Nothing can truly be understood. I called this "The Realm of No Definitions", and it is the condition of our being.

I came to see ideas as mere decorations. Not much different from decorations on a Christmas tree, or how one might decorate their house. But it was more than that. Ideas are not just mere decorations. They can have practical utility too. But seeing this practical utility and how to utilize ideas for practical purposes took on a purely aesthetic quality. It was art, pure and simple. One could surround themselves with beautiful, useful ideas and this is like making a good piece of art. It is the art of living one's life. It is the art of decorating life with beautiful, interesting, helpful, and useful ideas.

Or, as you see in the world around you, most people are unaware completely of this level of thinking and are nothing but slaves to the ideas in their minds. These people are in no position to be artistic with their ideas and their mind. No, they are just a bunch of deluded gumbas, who think the world is this or that, and act in accord to how the ideas drive them, making them just a bunch of automatons.

But they only seem like automatons. They are not automatons. They are people and they are either dishonest, or afraid, or ignorant, or some mixture thereof. The ideas are just the surface, and under the surface are strange emotional twists and unfulfilled desires. And there is stuff below that too. The hidden aspects of the mind go very deep, once one gets up the courage to peer down there (or is stupid enough to take 5 hits of acid and get violently thrown down there).

The apparent automaton thing is true of people that believe in a specific religion, or people that believe in science, or people that believe in some social cause, or some specific philosophy, or are liberal, or conservative. Whatever. The list of shit people believe is endless. All compassion aside, one and all they are just a bunch of self-deluded idiots because they don't see how they are letting ideas run their experience, instead of putting ideas in their rightful place as just another element in their experience; an echo, a reflection, a decoration.

People worry about the clothes they wear, but they don't worry about the ideas they use to decorate their lives. Idiots, one and all.

So, given that this is the main way I have looked at the world for the past 30 years, you can imagine my shock, surprise, and chagrin to read this from the great Swami Krishnananda:

> "When the ultimate cause of a particular experience is discovered, it will be found that the cause lies in the recognition of the Self in the not-Self. This was the definition of avidya given by Patanjali"

"Avidya" is the yoga word for "ignorance". Not just ignorance that can be erased by learning some information. No, avidya is a type of ignorance for which there is no translation in English. It is a cosmic ignorance. It is a wrongness of such epic proportions that it creates universes, life, and consciousness as we know it.

Avidya creates our experience...

ℒ Experience: Interlude

I mentioned in the previous post that my mind spun and spun while under the influence of five hits of acid. This type of experience is almost impossible to express in words. But there is one person I know who had the writing skills, the imagination, the word-smithing creativity to capture at least some facet of what such an experience is like.

As an interlude to this essay on experience, I wish to recall his words, to expose the Reader who may be unfamiliar with such things some small flavor of what was going through my mind that night. While these are his thoughts, mine were similar; different yes, but similar in essence.

This long quote is from Alan Watts' wonderful book *The Joyous Cosmology*. For those who have never read this book, it is one of the most insightful books one can read. Note that a smaller quote below will be the basis for Chapter 3, the paragraph beginning, "Life seems to resolve itself down…" We will come back to this thought next time.

So….Without further ado, Ladies and Gentlemen it is my pleasure to introduce Alan Watts:

> "I am listening to a priest chanting the Mass and a choir of nuns responding. His mature, cultivated voice rings with the serene authority of the One, Holy, Catholic, and Apostolic Church, of the Faith once and for all delivered to the saints, and the nuns respond, naively it seems, with childlike, utterly innocent

devotion. But, listening again, I can hear the priest "putting on" his voice, hear the inflated, pompous balloon, the studiedly unctuous tones of a master deceptionist who has the poor little nuns, kneeling in their stalls, completely cowed. Listen deeper. The nuns are not cowed at all. They are playing possum. With just a little stiffening, the limp gesture of bowing turns into the gesture of the closing claw. With too few men to go around, the nuns know what is good for them: how to bend and survive.

ALAN WATTS

America's most adventurous philosopher describes his own experiences, ranging from the diabolic to the divine, with the "mystic drugs"—LSD-25, mescalin, and the mushroom derivatives

THE JOYOUS COSMOLOGY
Adventures in the Chemistry of Consciousness

By ALAN W. WATTS Foreword by Timothy Leary and Richard Alpert, Center for Research in Personality, Harvard University. Illustrated, $5.00, now at your bookstore.
PANTHEON

But this profoundly cynical view of things is only an intermediate stage. I begin to congratulate the priest on his gamesmanship, on the sheer courage of being able to put up such a performance of authority when he knows precisely nothing. Perhaps there is no other knowing than the mere competence of the act. If, at the heart of one's being, there is no real self to which one ought to be true, sincerity is simply nerve; it lies in the unabashed vigor of the pretense.

But pretense is only pretense when it is assumed that the act is not true to the agent. Find the agent. In the priest's voice I hear down at the root the primordial howl of the beast in the jungle, but it has been inflected, complicated, refined, and textured with centuries of culture. Every new twist, every additional subtlety, was a fresh gambit in the game of making the original howl more effective. At first, crude and unconcealed, the cry for food or mate, or just noise for the fun of it, making the rocks echo. Then rhythm to enchant. Then changes of tone to plead or threaten. Then words to specify the need, to promise and bargain. And then, much later, the gambits of indirection. The feminine stratagem of stooping to conquer, the claim to superior worth in renouncing the world for the spirit, the cunning of weakness proving stronger than the might of muscle—and the meek inheriting the earth.

As I listen, then, I can hear in that one voice the simultaneous presence of all the levels of man's history, as of all the stages of life before man. Every step in the game becomes as clear as the rings in a severed tree. But this is an ascending hierarchy of maneuvers, of stratagems capping stratagems, all symbolized in the overlays of refinement beneath which the original howl is still sounding. Sometimes the howl shifts from the mating call of the adult animal to the helpless crying of the baby, and I feel all man's music—its pomp and circumstance, its gaiety, its awe, its confident solemnity—as just so much complication and concealment of baby wailing for mother. And as I want to cry with pity, I know I am sorry for myself. I, as an adult, am also back there alone in the dark, just as the primordial howl is still present beneath the sublime modulations of the chant.

You poor baby! And yet—you selfish little bastard! As I try to find the agent behind the act, the motivating force at the bottom of the whole thing, I seem to see only an endless ambivalence. Behind the mask of love I find my innate selfishness. What a predicament I am in if someone asks, "Do you really love me?" I can't say yes without saying no, for the only answer that will really satisfy is, "Yes, I love you so much I could eat you! My love for you is identical with my love for myself. I love you with the purest selfishness." No one wants to be loved out of a sense of duty.

So I will be very frank. "Yes, I am pure, selfish desire and I love you because you make me feel wonderful—at any rate for the time being." But then I begin to wonder whether there isn't something a bit cunning in this frankness. It is big of me to be so sincere, to make a play for her by not pretending to be more than I am—unlike the other guys who say they love her for herself. I see that there is always something insincere about trying to be sincere, as if I were to say openly, "The statement that I am now making is a lie." There seems to be something phony about every attempt to define myself, to be totally honest. The trouble is that I can't see the back, much less the inside, of my head. I can't be honest because I don't fully know what I am. Consciousness peers out from a center which it cannot see—and that is the root of the matter.

Life seems to resolve itself down to a tiny germ or nipple of sensitivity. I call it the Eenie-Weenie—a squiggling little nucleus that is trying to make love to itself and can never quite get there. The whole fabulous complexity of vegetable and animal life, as of human civilization, is just a colossal elaboration of the Eenie-Weenie trying to make the Eenie-Weenie. I am in love with myself, but cannot seek myself without hiding myself. As I pursue my own tail, it runs away from me. Does the amoeba split itself in two in an attempt to solve this problem?

I try to go deeper, sinking thought and feeling down and down to their ultimate beginnings. What do I mean by loving myself? In what form do I know myself? Always, it seems, in the form of something other, something strange. The landscape I am watching is also a state of myself, of the neurons in my head. I feel the rock in my hand in terms of my own fingers. And nothing is stranger than my own body—the sensation of the pulse, the eye seen through a magnifying glass in the mirror, the shock of realizing that oneself is something in the external world. At root, there is simply no way of separating self from other, self-love from other-love. All knowledge of self is knowledge of other, and all knowledge of other knowledge of self. I begin to see that self and other, the familiar and the strange, the internal and the external, the predictable and the unpredictable imply each other. One is seek and the other is hide, and the more I become aware of their implying each other, the more I feel them to be one with each other. I become curiously affectionate and intimate with all that seemed alien. In the features of everything foreign, threatening, terrifying, incomprehensible, and remote I begin to recognize myself. Yet this is a "myself" which I seem to be remembering from long, long ago—not at all my empirical ego of yesterday, not my specious personality.

The "myself" which I am beginning to recognize, which I had forgotten but actually know better than anything else, goes far back beyond my childhood, beyond the time when adults confused me and tried to tell me that I was someone else; when, because they were bigger and stronger, they could terrify me with their imaginary fears and bewilder and outface me in the complicated game that I had not yet learned. (The sadism of the teacher explaining the game and yet having to prove his superiority in it.) Long before all that, long before I was an embryo in my mother's womb, there looms the ever-so-familiar stranger, the everything not me, which I recognize, with a joy immeasurably more intense than a meeting of lovers separated by centuries, to be my original self. The good old sonofabitch who got me involved in this whole game."

Chasing tail

3. Experience the Eenie-Weenie

A t the heart of anything and everything is a weird kind of spiral movement that is like a dog chasing its tail...

"Life seems to resolve itself down to a ... squiggling little nucleus that is trying to make love to itself and can never quite get there."

Like a dog chasing its tail. What is the meaning of life? A dog chasing its tail. What is life made of? What is anything and everything made of? Something that reminds one of a dog chasing its tail.

My preoccupation with the idea of a Möbius spinning stems directly from insights equivalent to Allan Watts' realization. It is a weird spinning motion that seems to dovetail right back into itself, and arise out of its own center. It closes back on itself, but not in a normal three dimensional fashion. It is not just a closed loop, or a spiral spinning around. It is like a spiral, but it folds back into itself in a non-three dimensional way. And it wells up from its own center.

This idea/insight/perception has implications on many levels. Let's name a few:

String theory. Modern physics has backed itself into a corner called "string theory". The idea is that electrons, photons and all the other micro- microscopic "-ons" are not points in space, but are extended objects like little strings. This is a contentious issue in physics. Not something I want to get into here (see my two favorite physics blogs 1, 2, for ongoing bickering). But the contingencies of modern physics seem to point in no other direction.

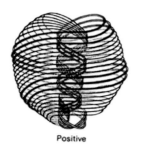
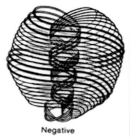
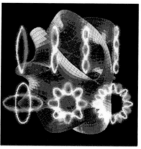

Positive   Negative

There is this stringy little something, spinning around, and it does so in a multi-dimensional space. This little squiggle thing that Alan Watts identified, the Eenie-Weenie, was at the base of everything, and might well be the true form of the strings. Just recently, the poor physicists have discovered that are some 10^{500} (that is a 1 followed by 500 zeroes for those who don't know scientific notation) possible forms for the stringy thingy. They don't have a way to tell which one is the right one. It's the one that is chasing its own tail and never quite getting there.

Kundalini. Kundalini is the metaphorical serpent at the base of the spine. Certain advanced yogic practices "awaken" this serpent. This is a metaphorical way to say that the practices activate a specific type of (extreme) experience in people's awareness. It is an extreme level of experience with much in common with the states of psychedelic inebriation discussed earlier.

The reason Kundalini is described as a coiled serpent is because, when it uncoils, one feels a spiral energy move up and down the main axis of the body, centered on the spinal column. At low intensity, it feels merely like shivers running up and down the back. At high intensity, it feels as if one is literally a slinky (you know, the toy) and that one's very being is a spiral energy or movement.

Again, this spiral movement is not a simple spiral or circular spinning motion. No, it has that weird

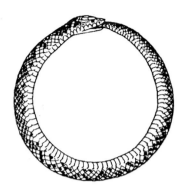

property of seeming to well up out of nowhere and to fold back into itself. Further, it's not just a sensation in the skin (called a somatosensation for you neurophysiology aficionados). The spiral sensation is also present in the mind. One's thoughts, emotions and perceptions also spiral in this weird self-eating fashion; the snake eating its own tail, Ouroboros.

Even if you don't believe in Kundalini, at the very least, this suggests that spiral patterns of electricity are whipping through the brain synaptic networks under these conditions.

Natural phenomena. How many things are spiral-like in nature? Hmmm? Galaxies: young ones. The old ones spiral themselves out and become big giant spheres. I mentioned the subatomic particles above: they all spin too. Solar systems are also spinning around. How many plants have a spiral shape? How about DNA and proteins? Their basic shapes are also spirals. Many of the fibers in animal bodies are spiral in shape: collagen, hair, skin, muscle. Even the pattern of how electricity moves through the heart can take on a spiral-like shape, spiraling around the chambers of the heart and affecting the contraction of the muscle.

Spiraling molecules combine together in ways that make other shapes. Hair appears straight but is made of such spiraling molecules. Sometimes the spirals manifest in the macroscopic form of the hair and people have "curly" hair to various extents. Millions of dollars are spent on products to straighten the hair out: an uphill battle against nature's spirals.

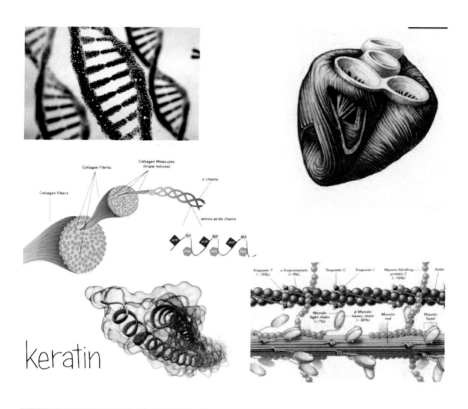

keratin

Spinning Out Of Control

I could keep going with such examples, but hopefully you get the point. Allan Watts' perception/insight that:

> "The whole fabulous complexity of vegetable and animal life, as of human civilization, is just a colossal elaboration of the Eenie-Weenie trying to make the Eenie-Weenie"

is not just some abstract thing. You can see it everywhere around you in living and so-called nonliving matter.

At this point, whether you think I am a raving lunatic or not, let us just grant that this Eenie-Weenie is at the root of things. That the root of things is nothing but a weird spinning motion, not just around and around in a circle, but weirder: it wells up from its center and spirals around and folds back into itself, trying to grab at its own center. But it disappears into itself before it can latch back onto itself: "As I pursue my own tail, it runs away from me."

What does this have to say about our life and experience? What are the implications?

When one asks: "what am I?", or "Who am I?", and probes deep enough, they discover the Eenie-Weenie. It is just a spinning motion. But who or what is doing the spinning? No matter how hard one peers into it, one always gets the same answer: no one, nothing. Nothing is doing the spinning. There is only spinning; round and round forever chasing its tail. Welling up from seeming nowhere and fading back into whence it came, and never quite achieving its goal of touching itself.

4 Experience: Mirage

"Every object in the world promises satisfaction, but it never gives satisfaction – it only promises".
- Swami Krishnananda

I n this series on experience, we've been exploring what happens when the curtain gets pulled back and the true identity of the Great and Powerful Oz gets revealed.

When you pull back the curtain, it's not ever what it seems to be…

When one comes to peer into the depths under the surface mind one sees patterns of emotional longings, unfulfilled desires, and a seemingly infinite list of wants.

Hollywood propaganda masquerades as common sense if you believe the meaning of life is to be happy. It is an absurdity of the first order to think we have (seemingly) only our brief life to try to extract out as much happiness as possible before we die and return to an eternity of oblivion and non-existence. Somehow, our life is supposed to follow the story structure of The Hero: we begin in some naïve Garden of Eden, then crisis pulls us out of our slumber of naivety and, against overwhelming odds, we overcome our demons and dragons, thereby acquiring the peace and happiness that is ours by right of conquest.

If you think like this I'm sorry to inform you it is pure bullshit from top to bottom.

When you pull back the curtain and see the frenzy, the typhoon, the flurry and torrents of desire and emotion whipping around in patterns not unlike those spiral patterns we discussed last time, it is exactly as Alan Watts described. Behind this wall of human craving, one finds deeper cravings on which those were built.

The politician's reach for power is nothing but the ever fearful insect mimicking the branches upon which it crawls. The will to build skyscrapers, to conquer nature with machines and technology, is but the wolf hunting the rabbit to sate its ever-recurring hunger. The desire for status and prestige is nothing but the mating ritual of dancing peacocks. The wish for happiness and security is the desire to regress to suckling Mother's warm, soft breasts. The craving to own things, anything, is like the drowning man's futile attempt to "catch even a straw that is floating on the surface of water" in the face of certain death. That includes owning ideas as much as owning material stuff.

We pull back the curtain of the surface of our mind to find only the second curtain of swirling human desires. We pull back the second curtain and we see the swirling desires of life itself, the mindless urge for survival, instinctive, unquestioning, and self-reproducing. We pull back this third curtain and see the swirling energies at the root of all things, making infinite patterns, making time, space, and energy. Time, space, and energy: the dog chasing its tail; as it pursues itself, it runs away from itself. That eternal futile effort is the base of space, time, energy, matter...life and stuff.

What is experience? What is life?

> "Every object in the world promises satisfaction, but it never gives satisfaction – it only promises".

It is promises. Only promises. Empty promises. Never satisfaction. To quote again Swami Krishnananda:

> "We love life very much; but it is not life that we love— rather, it is the pleasure of life that we love. If it was all horror and death-like pangs, one would not love life. But there is a drop of honey mixed with the venom of tense activity, and one is after

the little drop that is sticking even to the blade of grass which can cut one's tongue— due to which, life is kept moving."

One is willing to stick it out only for the brief taste of that drop of honey. And when you finally taste the drop of honey sticking to the blade of grass, your tongue gets cut. Regarding the taste of the honey once acquired, David Bowie informs us:

"Every time I thought I'd got it made, it seemed the taste was not so sweet"

What do you call an empty promise, a promise that only promises but never fulfills? Is it a lie? A deception? An untruth? Those words imply a type of intentionality that doesn't capture the scope of the situation. For the moment, perhaps the best term is to call it a mirage. It looks like something is there, when really, nothing is there.

We run on mirages as a car runs on gas. Mirages in our minds, drawing us forward through time and space, ever shifting, ever changing mirages: arbitrary, patterned, an ever-transforming kaleidoscope of mirages. The instant we think we have grasped it, it transforms into something else much different than what we were chasing after in the first place. And on it goes, on and on and on and on it goes.

In this mirage is the truth of all philosophy. It is in our minds, as the Idealists tell us. It is the World, the Universe and Stuff as the Materialists tell us. The Dualists are right too because it is both mind and matter. The Reductionists are also right: it all reduces to Ennie Weenies who can never catch themselves. It is meaningless and absurd as the Existentialists teach us, because it is always changing and ultimately makes no sense. But although it is always changing, it is patterned, and at any given moment appears to have order and logic, as the Rationalist tells us. There is reason in the temporary pattern, but it is not the reason the Rationalists hope it will be.

The reason is the patterns of desires, of longings, of hope, and it makes sense only by its consistent lack of fulfillment. It is always holding out the sign: "come hither for your satisfaction". When you get there you see in the distance another sign that reads, "come hither for your satisfaction".

And so it goes, eternally, round and round.

This is not postmodern anomie that is being described here, nor existentialist angst nor even nihilism. No, not by a long shot. These are but specific, and dare I say, petty, manifestations of a much more comprehensive condition. What is being described here is cosmic in scope.

The Buddhists call it Saṃsāra, the eternal wheel of life, death and rebirth. The Hindus call it Maya, the inscrutable; the causeless cause, without reason, explanation or purpose. The West calls it "life" and loves it and embraces it and glorifies it, but also, hypocritically, kills it, exploits it, and is blindly and ignorantly driven by it. The latter state of mind is called Avidya. It is the condition of our becoming: beckoning with promises, only to be rewarded with more promises, never fulfillment.

When one sees the ever-spinning, ever-changing kaleidoscope of mirages within mirages within mirages for what it is, there is only one rational response:

MAKE IT STOP!!!!!!!

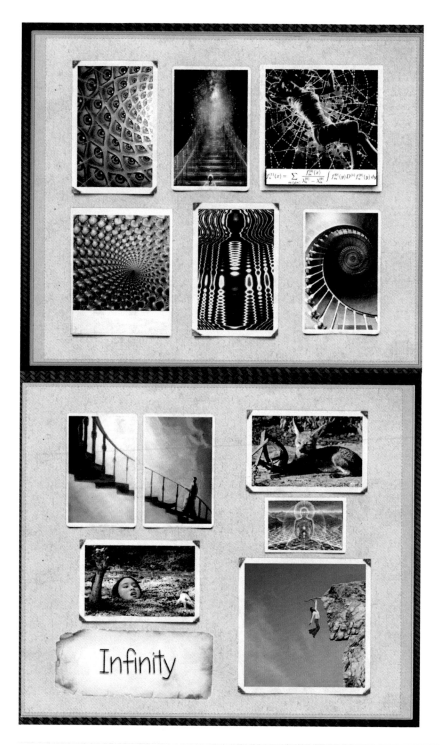

Infinity

5 Experience: The Infinite

When, through whatever means – yoga, psychedelics, dreaming, personal crisis and tragedy, getting hit on the head – one comes to see through the mirages of one's own mind, one becomes open to forces from which those still deluded are protected. Those forces must be understood to avoid getting lost in the infinity of mirages.

The delusions of those who believe in something, in anything, coupled with the inability to see below the surface of the mind, serve as a protective cocoon, or a womb, that shields such souls from the greater psychic ocean in which all of us are embedded. This psychic ocean was described last time in terms of curtains of swirling energies. Beneath the surface conscious mind is the world of human desires. Beneath this is the level of the blind instinctive urges of life. Beneath this is the torrent of seemingly infinite patterns that make up the Cosmos.

In reality, these are not curtains in any sense. They are more akin to vast jungles of psychic states, and they are like spider's webs and quicksand to one who does not understand them. Here we want to discuss the deeper delusions that are possible when these forces enter one's conscious mind. We begin by contrasting this condition to the state of mind most would consider "normal".

"Normal" People

So-called "normal people" are those souls still caught in the womb of their surface mind. We recall aphorism 2-4 of the Shiva Sutras:

४. गर्भे चित्तविकासोऽविशिष्टविद्यास्वप्नः ।

garbhe cittavikāso 'viśiṣṭa-vidyā-svapnaḥ

Which I.K. Taimni transliterates as:

गर्भे within the womb in which the development of the manifested universe takes place, i.e. *Maya* or *Prakrti* चित्त mind विकासः developed, unfolded अविशिष्ट lower, not the highest विद्या knowledge स्वप्नः is the nature of a dream, is imaginary and therefore unreal.

And translates as:

> II-4. "The lower kind of knowledge which develops through the mind in the realm of Maya or Prakrti is of the nature of a dream, i.e., purely imaginary and not real."

It thus is a recurring theme in yoga to associate the "normal" state of consciousness as ontologically akin to a dream. I quoted Krishnananda previously:

> "...when we wake up from world-consciousness. All these wonders, attractions and repulsions, these horrors, these forms of ugliness, these mysteries – all will be wiped out in a second when this relativity-consciousness gets sublimated in Absolute-consciousness, which is similar to the mind waking up from a dream into this world- consciousness..."

This leads to a view of Humanity different from mainstream Western views. In the yogic view, the "normal" mind, the mind confined to its own surface, has much in common with pollen grains on the surface of water. The pollen is buffeted in all random directions by a process called Brownian motion. Similarly, the surface mind of the "normal" person is buffeted about by life in all random directions by the forces underneath the surface mind. Such people laugh through the ups and cry through the downs of life, and all of this is taken at face value.

Certainly one knows their own emotions, "gut feelings" (intuitions) and reactions. But it is also true that one does not know from whence these arise. From van der Leeuw:

> "Our very consciousness is terra incognita; we know not the working of our own mind. What is it that happens when we think or feel, when a moral struggle takes place in us, when we are inspired, respond to beauty or sacrifice ourselves for others? It is as if we were prisoners in the vast palace of our consciousness, living confined to a small and bare room beyond which stretch the many apartments of our inner world, into which we never penetrate, but one of which mysterious visitors – feelings, thoughts, ideas and suggestions, desires and passions – come and pass through our prison, without our knowing hence they come or whiter they go."

Only rarely may faint and nebulous suspicions arise of the swirling forces underneath. But never is this clearly grasped for the "normal" person, never is the curtain pulled back. The immense and abstract psychic ocean in which we are all immersed remains the invisible background of what is known. Those confined to the surface mind are but puppets of the swirling abstract forces from whence the human condition arises.

This level of darkness and ignorance is a level all souls go through. Such people play the vast majority of roles in war and peace, in innovation and art, in the rise and fall of civilizations. As such, this level of existence is no different than any other form in nature: a tree, a grain of sand, a sun beam, a blade of grass, the planets, the other living creatures, a rain drop, solar systems, galaxies: and humans who live on the surface of the mind.

One cannot pass judgment on the mind in this condition because it is an eternal level in the scheme of things. Is a grain of sand, a rain drop, a tree inherently good or bad? Is a planet in its majestic orbit about a star good or bad? Is a tornado or super nova good or bad? Is day good and night bad? Assigning an inherent moral value to the things of nature simply makes no sense. Similarly, Humanity cannot be judged. It is a thing of nature, a part of the Maya, and is neither good nor bad in and of itself, but simply is what it is.

Looking Under the Hood

But once one's mind gets blown back, when, for whatever reason, by whatever cause, the deeper layers flood into the conscious mind, the conditions become irreversibly altered. The so-called "normal people" have no idea of these other states of consciousness, of how literal they are, of how abstract they are. Of how they cast a perspective on the so-called "normal state", how they provide contrast and insight.

Even if the curtain was blown back only briefly, giving but a glimpse of some other states of mind, this is enough to lift one out of the complacency of "normalcy". It is the contrast that is important. When one knows only a single condition, one is completely blind to that condition. It is only by contrast to another state that the so-called "normal state" can be recognized as such.

Then we must consider the conditions under which the curtain is pulled back. Is one prepared, and if so, to what degree? Or is it an unprepared exposure to things for which our language has no words and our culture has no concepts? In either case, a whole new level of danger comes into play and the relative degree of preparedness can help to blunt the impacts of these new dangers.

The Intermediate Zone

Sri Aurobindo offers one of the most cogent analyses of the dangers facing the person for whom the curtain has been pulled back, to whatever degree. He termed the inner layers The Intermediate Zone. The Reader is encouraged to read Aurobindo's words. To quote Alan Kazlev, keeper of the Kheper site, the Intermediate Zone "is a beguiling and dangerous transitional region or stage or whole series of stages between ordinary consciousness and complete enlightenment."

In the Biblical Old Testament, this concept is captured by the idea of "False Idols", the ones Moses broke with such vehemence. In the New Testament, the notion is captured in The Temptation of Christ by Satan. The Christian parables, however, are metaphorical, though what they allude to is quite literal. In Patanjali's *Yoga Sutras*, the Intermediate Zone is described as the "siddhis", the "super powers" that result from practicing samadhi, which are but impediments to enlightenment.

My favorite aphorism pertains to this issue:

५२. स्थान्युपनिमन्त्रणे सङ्गस्मयाकरणं पुनरनिष्ट-प्रसङ्गात् ।

Sthāny-upanimantrane sanga-smayā-karaṇaṃ punar anista-prasangāt.

स्थानि[न्] (by) the local authority; the superphysical entity in charge of the world or plane; powers of spaces उपनिमन्त्रणे on being invited सङ्ग attachment; pleasure स्मय wonder; pride; smile of complacence अकरण avoidance; no action of पुन: again अनिष्ट (of) the undesirable; the evil प्रसंगात् because of the recurrence or revival.

We use <u>Swami J's translation</u>:

> "3.52 When invited by the celestial beings, no cause should be allowed to arise in the mind that would allow either acceptance of the offer, or the smile of pride from receiving the invitation, because to allow such thoughts to arise again might create the possibility of repeating undesirable thoughts and actions."

This is my favorite aphorism for a couple reasons. First, the warning itself is important for those who have had the curtain pulled back. Second, it indicates the depth of what is at issue, yet in a slightly humorous way, although I am certain Patanjali was not trying to be funny. Still, the idea of being warned about being taken notice of by great Celestial Beings strikes my funny bone.

The Christian and Hindu ideas above are mutually consistent and the message is clear: when the curtain is pulled back, when one dips below the surface mind, one is now open to a whole new gamut of forces of distraction, and the chances of ensnarement have increased a million-fold compared to the so-called "normal" person who is merely buffeted unconsciously by such forces.

Getting Lost In Infinity

Aurobindo speaks of the incomplete nature of Intermediate Zone experiences:

"...idea-truths may have come down into him are partial only..."

He says:

> "These things, when they pour down or come in, present themselves with a great force, a vivid sense of inspiration or illumination, much sensation of light and joy, an impression of widening and power. The sadhak feels himself freed from the normal limits, projected into a wonderful new world of experience, filled and enlarged and exalted; what comes associates itself, besides, with his aspirations, ambitions, notions of spiritual fulfilment and yogic siddhi; it is represented even as itself that realisation and fulfilment. Very easily he is carried away by the splendour and the rush, (but) ... experience is usually lacking which would tell him that this is only a very uncertain and mixed beginning."

Aurobindo speaks in a personal way, about how these forces impact the personality. This is an important way to convey the facts because it is how we encounter them, and his advice provides a guidepost for coping. The essence is that the new experience seems so grand, so expansive compared to our normal, mundane everyday experience. But this increase in relative greatness is itself the clue that what is experienced is only but a partial truth.

In the classical Raja Yoga, the main emphasis is on **abhyasa** and **vairagya**: practice and dispassion. These are the time tested methods used to protect and prepare the aspirant for when the curtain gets pulled back. Abhyasa are the prescribed practices of yoga that lead to nirodhah or restraint of the mind. Vairagya, dispassion, is the attitude required as the effects of the practices manifest.

The results of abhyasa are the siddhis, the "super powers" that come spontaneously from silencing the mind. They give rise to the expansive feelings of being that Aurobindo discusses. What is described clinically and systematically by Patanjali, Aurobindo describes from the perspective of the person becoming drunk with only partial knowledge, partial truth, which is mistaken for complete truth because of a failure to appreciate the relative nature of the partial truth.

That is the key to understanding these warnings: they are saying that a relative truth, no matter how grand, how expansive, is still but a relative truth.

Such relative truths seem to unfold without end. If one avoids the pitfall of getting entangled in a particular such truth, then one has the chance to discover the seemingly endless staircase of relative truth, of the ever-unfolding landscapes of ever-widening realization and being.

One ascends this staircase: through the forces of the personality, beyond the conscious mind, through the bizarre unconscious imagery beyond which are the forces of Humanity and its intrigues and dramas, into the forces of life on the Earth, of all the creatures and plants and the energies that unite them into one giant network. Beyond this are the forces of the Earth and its planetary kin as members of the solar system, where the Sun's energies dominate and control the countless patterns of revolution and events in our solar system. Beyond this are the forces of our Galaxy, the Milky Way, where forces and abstraction are such that the Sun is but a grain of sand, whisked along in a torrent beyond any human comprehension. And thence to the extra-galactic forces, and beyond this, cosmic realities that hold in their metaphorical hands the power of universes. And beyond this are universes within universes within universes and the staircase of greatness fades into a fuzzy horizon where invisible glories can only be guessed.

The picture of an infinitely ascending staircase conveys the point: it seems to go on forever and ever and ever: greater truths encompassing lesser truths, greater beings encompassing lesser beings, glory beyond glory. Never ending, ascending seemingly to infinity.

And then we find ourselves lost in infinity, ensnared in the web with no end, falling down the proverbial bottomless pit. We become the condition of Prometheus who, trapped in bondage, had his liver torn from him each day, only to have it grow back so it could be ripped from him again the next day.

The illusion is infinity itself. Each level no different than our original insight:

"It is always holding out the sign: "come hither for your

satisfaction." When you get there you see in the distance another sign that reads "come hither for your satisfaction".

That is really what the myth of Prometheus conveys: the illusion of infinity and the infinite disappointment of being incomplete every day. The pain inflicted by the Eagle ripping out his liver is symbolic of the condition of incompleteness that accompanies relative being. Prometheus thought he was stealing fire from the Gods. No, he was just deluding himself by getting trapped in infinity and finding that each new level (each new day) brought the same pain of incompleteness.

While symbolism is fun, I prefer plain talk such as given by Patanjali et al. For those for whom the curtain has been pulled back, when the warnings of such as Patanjali and Aurobindo are not heeded, we fall from the frying pan into the fire. We get lost in infinity.

PLANETALK
ABOUT THIS AND THE OTHER
WORLDS

LEARN

ABOUT THE OTHER WORLDS…

PLANETALK ARCHIVES

PAST BLOG POSTS

SEPTEMBER 2014 (2)

AUGUST 2014 (2)

JULY 2014 (3)

JUNE 2014 (2)

MAY 2014 (8)

APRIL 2014 (5)

MARCH 2014 (1)

JANUARY 2014 (1)

NOVEMBER 2013 (5)

Follow PlaneTalk

BLOGS I FOLLOW

1. Psychiotron
2. The World Knot
3. The World As Computation
4. Baldscientist
5. onehindu
6. Looking at Life

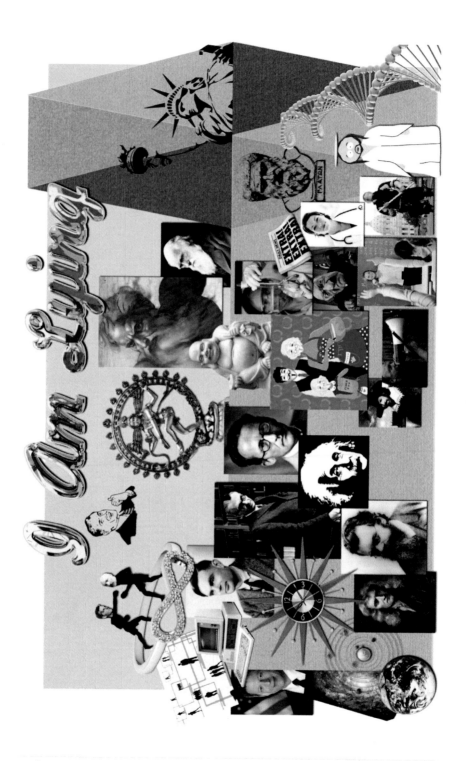

6. Experience: Nothing

R elativity is incompleteness. This condition comes about through a projection of our being seemingly outside of ourselves. But this is an impossible act, and the result is mirage, illusion, nothing.

The Pride of Reason

This is part 6 of the discussion on experience. As most people know, 7 is considered a "lucky" number: there are 7 colors in the rainbow, 7 notes to a major scale, 7 days in the week, 7 planes of nature, God created the world in 7 days, etc. etc. 7 is consider the number of wholeness and completion.

The number preceding 7, 6, is therefore associated with incompleteness, the state of "not quite there yet". That is our topic now, to build on the previous observation that infinity, in the sense of a never-ending progression, is in fact the very embodiment of, perhaps the purest expression of incompleteness. Infinity is not a thing. It promises to be a thing, but never is.

I did not plan that Chapter 6 would be about incompleteness, but it is, and I find myself coming back to Chapter 6 of Alister Crowley's *Book of Lies*, which I wrote about many years ago in Chapter 16 of _Beyond The Physical_. Crowley's chapter is so short it can be reproduced here in full:

"The Word was uttered: the One exploded into one thousand
    million worlds.
Each world contained a thousand million spheres.
Each sphere contained a thousand million planes.
Each plane contained a thousand million stars.
Each star contained a many thousand million things.
Of these the reasoner took six, and, preening, said: This is the
    One and the All.
These six the Adept harmonised, and said: This is the Heart of
    the One and the All.
These six were destroyed by the Master of the Temple; and he
    spake not.
The Ash thereof was burnt up by the Magus into The Word.
Of all this did the Ipsissimus know Nothing."

Crowley's commentary contains one of my all-time favorite lines:

"The Rationalist takes the six ...and declares them to be the
universe. This folly is due to the pride of reason."

Taking six things out of the gazillions of things that exist and declaring
them to be everything could almost serve as the psychiatric definition of
insanity in the DSM. Anyway, I wrote about this in *Beyond the Physical*
years ago and won't traverse that ground again here.

The point of citing Crowley here is that, as you see, the number 6 is
portrayed as the infinity of things that make up the manifested universe.
These were discussed last time as a never-ending progression of forms,
all of which are characterized by the fact that they promise, but never
give, fulfillment.

We now consider a less poetic and a more intellectual exposition of
infinity.

Infinity as Math Object

There is much written on the topic of the flirtations of Western
civilization with the concept of infinity. Well-known stories begin at the
ancient Greeks, wind their way through Nicholas of Cusa, Saint
Augustine, Galileo, Leibniz, and many, many others, and culminate in

the work of Georg Cantor, from whence a downhill progression brings us to now.

The story often starts at Aristotle's distinction between potential and actual infinity. The counting numbers, 1, 2, 3... are potential infinity. They imply infinity as a potential state (notated by the "..."), but one never gets to infinity. Instead, one seems to be able to get as close as one wishes by just counting higher and higher. However, the end, infinity, is never reached.

Further, the idea that one gets closer to infinity by increasing numbers is purely illusory. For no matter how high we count, we can then double this, multiple by 10, by 1000, by 1,000,000 and we are no closer to infinity than when we were at 1, 2, 3... . As Cantor himself showed, no matter what magnitude of number we consider, that magnitude is exactly zero next to infinity. This is to say, the concept of "magnitude" makes no sense in the context of infinity.

Actual infinity is the idea that we can hold something infinite in the finite palm of our hand. Cantor was the first mathematician to formalize how this could happen. This led to paradoxes that made Russell look like the intellectual amateur he was, and generated the foundations of mathematics, which Wittgenstein thought were a complete travesty of the intellect. Wittgenstein, being a philosopher and not a mathematician, was mostly ignored, and people like Gödel and Turing did their paradoxical tricks showing how, with mathematical certainty, we can define incompleteness and unknowing, based on ideas analogous to Cantor's.

All of this led to computers, which led to the widespread appreciation of chaos theory and fractals, which previously existed only in the minds of "special people" like Poincare, Cantor, and a handful of others.

Now, we can indeed hold a simple equation in our hand, like

$$z_{n+1} = z_n^2 + c$$

which is the equation of the Mandelbrot set, an infinite entity we can, so to speak, hold in the palm of our hand, or at least grasp in our mind as an infinite thing that occurs in a finite space.

The main reason for the above is to lay out a little about Western Civilization's notions of infinity so we can contrast them to Hindu ideas of infinity. They are similar, but different. In Hindu thought, a distinction like potential and actual infinity also exists. But it is interpreted in a different fashion. Each are given a status as real or unreal, but in the opposite way from the Western ideas.

The Measure of All Things

To the West, potential infinity is real because we can write down equations of all types for which we can imagine what would happen at infinity, even if, by any practical means we cannot get there. We can deduce (or I should say "intuit" since we cannot fully calculate it) what would happen at infinity, which is the mathematical idea of a limit. We then say an equation "converges" to a finite solution, even though, in principle, an infinite amount of calculation would be required to actually compute the limit. For readers who have never seen a limit before, I've prepared a small example (Appendix 1: Example of a Limit) to illustrate the concept.

The idea of a limit is important. It, perhaps more than any other, is responsible for the technological world we now inhabit.

In this sense, potential infinity is real to the Western mind, but actual infinity does not exist, except in the imagination as a limit of a potential infinity, or in the imaginations of Cantor's followers.[1]

Contrary to the Western notions, to the Hindu mind, actual infinity is

---

[1] For the mathematically sophisticated: I'm going with Wittgenstein and Leopold Kronecker and treating Cantor's "Paradise" as a delusion. Writing a finite symbol and saying it is infinite is nonsense. No practical procedure can ever realize infinity. I agree with Wittgenstein that the diagonal method is flawed. Somehow this all links back to the real numbers. But not being a professional mathematician, I am in no position to technically argue the point. However, Poincaré in _Science and Hypothesis_ speaks of mathematical thinking as "reasoning by recurrence". Poincare's sophistication on this matter led him to reject Cantor. Demonstrably more practical good came from Poincaré than Cantor. In addition, the Hindu ideas are quite clear there is only one infinite (Brahman). Some further elaboration is provided in Chapter 7 but a full exposition will not be presented and may serve as a topic for a future post.

literally real and potential infinity is literally unreal.

Actual infinity is called **Brahman**, and it is spoken of in terms of what it is not. Brahman is not any finite attribute one can imagine. Therefore, even though Brahman is real, in fact the only reality there is, it is not something a finite being can experience or comprehend.

The idea is not confined to Hindus. Nicholas of Cusa described the situation with great beauty and penetrating insight and you should read what he had to say about "Learned Ignorance".

To the Hindu mind, potential infinity is similar to the Western idea of potential infinity as something that goes on forever: a thing that cannot be thought of, or described, or calculated in a finite way.

In the simplest of terms, any approach to infinity, any potential infinity, is always a process or procedure that can, in principle, repeat forever, like adding 1 to the previous number to generate the counting numbers.

This is the well-known idea of an *algorithm*: a series of well-defined steps that generates an output. Some algorithms have a finite number of steps. Others, such as counting by 1, can go on forever. The algorithm is the real thing (by "real" I mean "it exists" and am not referring to real numbers). If an algorithm repeats forever, its output, some potential infinity, is also real. But the actual infinity implied by the algorithm is not real; it is not something that can be made to exist.

But while potential infinity is real for Westerners, it is the very definition of unreal for the Hindu mind. The Hindu word for potential infinity is "Maya", which is translated in a variety of ways: magic, illusion, Mother, measure, ma = "man", the measure of all things.

The latter, "measure", from the root "ma-"(the source of the word "man"), is perhaps the clearest because it implies the essence of the idea: that which is relative. Where Brahman is absolute, Maya is relative.

The Relative

Man is the measure of all things. Things are bigger or smaller than us;

faster or slower than we move, heavier or lighter than we can lift, brighter or darker than we can see, louder or quieter than we can hear, and so on. "Ma", "man" "measure", "maya".

Relative means that the thing only makes sense in comparison to something else of like quality but different quantity. We cannot know what "bigger" means unless there is something of a different size. We cannot know what is good unless we know evil. We cannot know what is intelligent unless we know what is ignorant.

The implications of the Hindu view are well-known to anyone who studies Hindu or yogic thinking, but they are generally unknown to those familiar with only the Western understanding of infinity, at least nowadays. People like Leibniz and Nicholas of Cusa, were Westerners with ideas very similar to the Hindu notions, but they are rarely read today, let alone taken seriously, and are exceptions in Western thinking. Had their ideas caught on, Western thought would today more resemble Hindu thinking.

The implication was alluded to previously and is now stated: **we only know by contrast**. If there is no contrast, then there is only blankness, a nothing.

A common sense example is how we move about in the gaseous atmosphere of the Earth. We look forward and see nothing between us and other objects. We feel no resistance to moving forward (only the downward pull of gravity). Therefore, we consider the space around us to be empty, though it is not, and we only need feel the wind to know something invisible is there buffeting us.

Of course it is common knowledge we live in the atmosphere, but in our day-to-day experience, we take the air for granted because it is the only condition we experience. We can, in a manner of speaking, subtract it out of our moment-by-moment considerations.

This same logic was used in Chapter 5 to describe "normal people" who only know the surface mind. Again, without the contrast of experiencing an altered state of consciousness, the "normal" person takes their conscious state for granted and makes the (incorrect) inference that it is all that exists (that pesky "pride of reason" thing).

The point is that what we call "knowledge" or "information" is an expression or manifestation of contrast. Contrast implies difference. Difference implies relative-ness. Therefore, all we sense, all we think, all we know, all we *experience* is relative. This is what Hindus call "Maya". Our entire existence is of a relative nature. We know only Maya.

This does not change when the curtain gets pulled back. The inner realms are also relative: *expanded* being, *greater* consciousness, *more* awareness. The adjectives are the giveaway.

A key point I wish to convey, one of the central insights of this essay, is the following:

When we recognize that existence always has the quality of being beckoned by promises that never fulfill, this has the same form, the same "shape" as "1, 2, 3...", particularly the "...". The "..." is the promise. It seems like it will give to us, like it will fulfill, but it only goads us on to something else. "..." is not a sign of fulfillment, it is but a road to a different promise, and then the cycle repeats.

In short, potential infinity is the nature of the relative, and it is always incomplete, and most important: this is the nature of experience.

All Conditions Each

Perhaps the most important implication of realizing that existence is relative is to realize that any one thing depends on everything else for its definition, for its being, for its existence.

To be more precise, given that all relative things are in a constant flux, it's not even correct to call relative existence "being". As Western philosophers have recognized for over two millennia, the state of things must instead be called "becoming". It is always in a process of transformation.

I will not dwell on this point here, but leave it to the Reader to consider how, when in the relative condition, any seemingly separate thing depends, ultimately, upon all other seemingly separate things.

The end conclusion is there is no such thing as a separate thing; only the appearance of separate things, due to ignorance of not seeing the total picture of all the connections, overt and covert.

From Whence the Relative

It doesn't take much mental power to understand and accept that all experience is relative. Once it is pointed out, once all the terms are defined and the picture drawn, it is all pretty obvious. Self-evident actually. Axiomatic. What requires mental power and effort is to think through the implications, particularly if we wish to stay honest about the matter.

For the honest truth is that nothing we can do in actual fact can make an actual infinity. A fractal is infinite in our imagination, not in fact. A convergent series is infinite in our imagination, not in actual fact. No one has ever written down an infinite set of symbols. (Please, if you are so smart, write me a true infinite set. I will be impressed and you will get a Fields Medal). No infinite calculation can be performed. No amount of "..." can change the fact. That was Wittgenstein's point: "..." is simply dishonest. It is a lie, a sham, and a scam. Be that as it may, it is not the point I want to dwell on here.

The issue to consider is: from whence has this condition arisen? We stated the answer in Chapter 1:

> "When the ultimate cause of a particular experience is discovered, it will be found that the cause lies in the recognition of the Self in the not-Self."

When the cause of a particular experience is discovered, the cause of all experience is discovered. Since all experience is relative, the cause of the relative condition is discovered.

The cause is found to be a process of projection, a trick, like a mirror image. It looks like the reflection is really something inside the mirror. But it is not. It is just a trick of how light reacts with mirror surfaces. Something analogous happens in consciousness. Consciousness makes a projection, then reflects itself in that projection, and takes it for a reality, when it is only a reflection.

Then, that which is doing the projecting begins chasing its projection, like a dog chasing its tail. The tail chasing process behaves like a potential infinity: it seems to go on forever and ever; it always promises but never delivers. From this tail-chasing process emerges what we in our ignorance call "reality", "existence", and "experience".

There are, at root, two main players in this Grand Illusion: being and becoming. Although I said all experience is becoming, behind the becoming is being, without which becoming would be impossible.

Borrowing a metaphor used by both Krishnananda and Taimni, the relationship between being and becoming is like watching a movie at the theater. The movie must be projected on a white screen. The white screen is the "invisible" background, invisible in the sense that it is not to be paid attention to while focusing on the move. But without it, there would be no movie.

This is the relationship between being and becoming. Being is the background, the substrate onto which becoming is projected. We experience becoming. But we only do so because we are. The actual infinity, which is being, projects itself and makes seemingly infinite potential infinities that spin round and round in the process of becoming, trying to achieve a satisfaction that can never come.

7. Experience: Life

"...they were in the habit of conferring honours among themselves on those who were quickest to observe the passing shadows..."

- If you don't recognize it, Google it and read the original

Covering Our Tracks

It behooves us at this point to gather our thoughts and review what has been said to this point. The following summarizes the main points intended to be conveyed by the previous parts of this essay:

1. Words and ideas cannot *capture* experience but they can (more or less perfectly) *reflect* experience.
2. Experience consists of always striving, but never quite getting there.
3. This always-striving seems to be present in all natural system we can observe.
4. Since experience consists in always-striving, images in our mind of completeness and fulfillment are just mirages.
5. Even the deeper layers of consciousness offer no solace. They do not improve the condition of always-striving found on the surface mind but only offer more of the same.
6. Always-striving reflects the incomplete nature of potential infinity. Manifest experience teases us with promises of completeness, but never delivers.

In short: we are driven forward by mirages of promises that never deliver, and then... we die.

Seemingly this is not the happiest of pictures. But it is honest. And honesty has its payoffs. For, by being brutally honest about the facts of our experience, we are now positioned to reap the rewards of that honesty, which is truth, real truth, not mirages.

That truth, elaborated below, goes something like this: Experience is but a vacuous mirage. As such it is nothing. However, this condition is like a mirror that shows an image when nothing is there to be reflected, which is impossible. The mirage of relative existence, experience, must be a reflection of something that actually exists, just as the image in the mirror must reflect something outside the mirror.

Other Skylines to Hold You

In this essay I have aped traditional yogic thought by construing A transforming into B as an urge for A not being fulfilled and, in the attempt at fulfillment, transforming into B. We can also construe such a transformation as an unpredictable creativity. The outcome of each process is effectively the same, but the overtones and implications of each view are different.

The first seems sad and pessimistic, the second seems optimistic. The optimistic view – creativity – blinds us from squarely recognizing our ignorance of the entire process. We may think of it as God, or as evolution, or human creativity, but in all cases, when the transformations of experience are seen as creativity, we must insert a nebulous idea of the cause of the creativity. Any such idea may make us feel good. But it bandages over the fact of our fundamental unknowing. As such, it is simply dishonest. As an act of dishonesty, it prevents us from seeing deeper truths, and so I dismiss such approaches here.

To make my point as clear as I can, let me state it in another way. In part 6 we discussed how Hindu and Western views of infinity are opposites:

1. Western view: Potential infinity is real, absolute infinity unreal (except to Cantor's dupes).
2. Hindu view: Potential infinity is the relative and is unreal, absolute infinity is the only reality.

We can make an analogous distinction:

1. Western view: Even though we are fundamentally ignorant, there is a creativity in experience (God, evolution, Man's imagination, whatever) that drives things forward.
2. Hindu view: We are fundamentally ignorant, and it is our ignorance that not only keeps things moving forward, but creates the illusion of forward motion.

When stated thus, the implications of each approach are clearer. In the first case, we forever chase our tail trying to find the source of the creativity. As such, it becomes just another thing chasing its tail.

Creativity

In the West today, there are two main views of creativity: evolution and God. I hardly wish to get embroiled in such a pedestrian debate where both sides are dominated by the most mediocre of minds[2]. The false core of all such debates goes back to the main point of Chapter 1: these people believe that ideas can take them beyond their own minds. Again putting all compassion aside, these people are one and all fools.

The smartest thing I have yet heard on this front is Stuart Kauffman's opinion. You can see it here[3]. Here I just paraphrase: because the results of evolution are fundamentally (e.g. in principle) unpredictable, the outcome looks the same as if God did it. So, if it quacks like a duck...

While Kauffman is trying to paper over the differences between secular humanism and religiosity, the real beauty of his ideas are that they get right to the heart of the matter. Kauffman penetrates past all the distracting and divisive issues that people whose religion is evolution very much enjoy fighting over with people whose religion is Christianity or Islam, etc.

The main point is that we will never know. We are fundamentally ignorant. The future is fundamentally unpredictable; the past, at best only partially knowable; the present is an inscrutable mystery.

---

[2] http://www.youtube.com/watch?v=Uaq6ORDx1C4
[3] http://www.youtube.com/watch?v=8I5mYDUARY4

So, it is just what I said above: whether you go the God or evolution route, you are just making a *"God of the gaps" argument* and pretending that you are not. As least the religious people know they are invoking God. Nonetheless, the whole spectacle is absurd.

So, while it seems so optimistic to speak of a "creative universe", whether God-driven or not, it all just bandages over our basic ignorance. "Creativity" becomes just a synonym for "I don't know". If it quacks like a duck...

The Eastern view actually is the optimistic view insofar as it suggests we may overcome our ignorance.

However, a cynic may turn my argument against me and accuse the Eastern view to be simply another example of chasing tail. It could be the case that the yogic view holds out just another promise, the promise of overcoming our ignorance, and as such is just another promise to chase to no end.

However, we now embark on showing that this is just not the case.

It's Worse Than We Thought

But before finding the light at the end of the tunnel, we need to go to the heart of the darkness and stare directly into our ignorance. The end of part 1 briefly mentioned this ignorance: avidya. There I stated:

> "It is a cosmic ignorance. It is a wrongness of such epic proportions that it creates universes, life, and consciousness as we know it."

We have now covered that ground, however imperfectly. Now we consider the cherry on the top, so to speak, the penultimate aspect of avidya, which, has been stated by many. The way Alan Watts said it in Chapter 2 is as good as any:

> "The trouble is that I can't see the back, much less the inside, of my head. I can't be honest because I don't fully know what I am. Consciousness peers out from a center which it cannot see—and that is the root of the matter."

And it really is the root of the matter. We cannot see our own eyes. We cannot jump out of our skin. We cannot think a thought that takes us outside of our mind. We seem to be trapped in a condition of permanent ignorance.

Drisimatrah

At the beginning, and throughout, we have alluded to the answer to this quandary. In fact, it was already spelled out as clear as it can be in Chapter 1:

> "When the ultimate cause of a particular experience is discovered, it will be found that the cause lies in the recognition of the Self in the not-Self."

The following ideas are not mine. They are Krishnananda's interpretation of some of the aphorisms of the *Yoga Sutras*. He spends many chapters in _The Study and Practice of Yoga_ explaining the concept and the interested Reader can get a fuller discussion there. Instead of trying to pick choice quotes, I will try my best to explain Krishnananda's main ideas.

I got at this before by the metaphor of the movie screen and the movie. The movie needs to be projected onto something, the screen. If the screen itself changed, the movie wouldn't work. The screen needs to be stable, unchanging, for the projection process to work.

This metaphor conveys a simple but accurate understanding of the Hindu and yogic ideas of how experience links to consciousness. Because "consciousness" is a word with many meanings in our culture, it is best to use the Sanskrit word here: **drisimatrah**. Drisimatrah is pure self-awareness, pure being, and pure bliss. It is called Sat-Chit-Ananda in Vedanta. Here we are interested in the Sat-Chit aspect: the aspect of being-awareness (this was also discussed in Chapter 7 of *What is Science?*).

Drisimatrah is neither something alien, nor metaphorical, nor intellectual, nor even difficult to understand. This is because we experience drisimatrah all the time. We can know it because sometimes we seem not to experience it.

We alternate between being awake and being asleep. When we sleep we either dream or we don't. When we have periods of complete unconsciousness, with no knowledge of anything, not even a dream, then this is the seeming lack of drisimatrah. When we are aware of anything, we experience drisimatrah. It is our consciousness, our self-awareness.

In our relative condition of becoming, drisimatrah waxes and wanes like the Moon. It has metaphorical day and night like the Earth. It is not turned on all the time. Why is it not on all the time? Because in the relative condition everything spins and moves in phases. As elaborated below, drisimatrah reflects its being in the relative condition, and thus *appears to* wax and wane like relative things do.

When drisimatrah is on, we are aware. Not only are we aware of things, we are aware that we are aware. Drisimatrah is the movie screen on which all of our experience is projected. More precisely, it is the medium in which all of our awareness of specific things, including our "self" occurs. As a medium, it is like a light in the sense that things become illuminated; the illumination provided by drisimatrah is our very awareness[4].

That is the first fact to appreciate in Krishnananda's account: what we call "awareness", drisimatrah, is the stable, unchanging ground upon which all of our specific perceptions and experience occurs. Most important, we all know drisimatrah, intimately. It is the ground of our first-hand experience.

Seeing the Self in the not-Self

The next step in Krishnananda's logic is this: the drisimatrah gets identified with the specific things occurring within it. Our awareness of being *is* drisimatrah. However, as he says in the quote above, drisimatrah projects its being onto the specific patterns that occur within it, that occur within awareness. In yoga, these patterns are called

---

[4] Side note: I don't want to get into a mind-matter debate here. I addressed this in *What is Science?* and am not going to repeat myself here.

"vrittis" the patterns in the mind. When vrittis are due to input from the senses, the stimulator of the vrittis is called "gunas".

As we have dwelled in the six previous chapters, the vrittis within awareness, within drisimatrah, and the gunas they reflect do not have being: they are always changing and therefore always becoming.

The drisimatrah which *is* imparts its "is-ness" to the vrittis. But the vrittis (and corresponding gunas if applicable) only seem to be, but really are not. The vrittis only seem to be because drisimatrah imparts its being to them.

This situation leads to a fundamental mix up of properties. The things that are always changing are given the property of being. The being, drisimatrah, assumes or takes on the form of the becomings—the transient flux of things—and misunderstands its own nature. The permanence of being is projected on to the transient fluxes, and the transient fluxes reflect back into the permanent being. This is the condition at the root of our experience. This condition permeates everything.

This is what Krishnananda means by "the recognition of the Self in the not-Self". The transient patterns become like a mirror, reflecting the being of consciousness, which cannot see itself as it *is* because it is distracted by all the patterns.

On one hand, it is good that consciousness reflects its being in the objects of perception because it lets consciousness know it exists. On the other hand, this becomes a problem. It becomes an impossible problem. Let's consider two levels where it is obviously a problem: (1) on an abstract philosophical level, and (2) on a personal level. We'll discuss each in turn.

The Fall From Grace

How can something be what it is not? How can something transient be permanent? How can something permanent be transient? Drisimatrah is the only permanent denizen of infinity. More precisely, drisimatrah *is* infinity, the absolute one, not the potential variety. Whether you believe this or not is not the point. Just grant the premise so we may continue with the argument.

On the other hand, all of the patterns, all of the vrittis that occur in the medium of drisimatrah are the Relative. They are transient, partial truths, impermanent and are of the form of potential infinities that, in their becoming, seem to be, but truly are not.

The act of drisimatrah projecting being onto the vrittis is the act of taking something transient as permanent. When the drisimatrah identifies with the vrittis (and the gunas reflected therein), it believes that the permanent is transient.

The very attempt to consider the transient permanent and consider the permanent transient is the essence, the core, of the "chasing tail" process.

It generates everything.

The logic now gets very abstract. For people familiar with Christianity, this is called the "Fall From Grace" and the process is much more abstract than the simple Bible allegories indicate. The necessity to consider such concepts is well beyond the intellectual capability of modern scientists and secular philosophers so let us just ignore their cat calls and derisive remarks and proceed forward.

Self-aware Being, drisimatrah, is all that exists. In its pure form it is called Brahman, and as such it is actual infinity. The act of projecting itself into what it can never be is called "Maya". Maya has no reason, no cause, no purpose. It is the causeless cause sought in all philosophy, and more recently by science. About all that can be said about Maya is that, in infinity, anything can happen, even the inscrutable act of the infinite trying to become finite. There is no cause, no beginning to this process. It happens outside of time and is eternal. The surface mind will simply never understand it. Deal with it and get over it because more important things will be required of your intellect as we proceed.

Infinity, Brahman, twisting itself into this impossible posture, tries to copy itself. But the copies are not identical copies, which is impossible (discussed below). Instead the copies are caricatures of infinity; they are the myriad potential infinities.

So it is a strange sequence. There is the absolute infinite, Brahman. There is some inscrutable facet of Brahman, Maya, by which Brahman tries to duplicate itself. But this is impossible and the result is a seemingly infinite set of potential infinities characterized by each trying to fold back on itself, eat its own tail, as a bizarre mimic of actual infinity. This strange, bizarre, inscrutable process is the "Fall From Grace": the descent of the Absolute into the Relative, the One into the Many, the Infinite into the Finite.

Why can't the potential infinities be real (i.e. absolute) infinities? Because absolute infinity is already all-encompassing. There cannot be two absolute infinities, let alone more.

Let us go all the way back to the dawn of the modern age, to 1440, to Nicolas of Cusa, who reflected deeply on this fact and saw the truth that has since been forgotten:

> 'Now, I give the name "Maximum" to that than which there cannot be anything greater. But fullness befits what is one. Thus, oneness—which is also being—coincides with Maximality. But if such oneness is altogether free from all relation and contraction, obviously nothing is opposed to it, since it is Absolute Maximality. Thus, the Maximum is the Absolute One which is all things. And all things are in the Maximum (for it is the Maximum); and since nothing is opposed to it, the Minimum likewise coincides with it, and hence the Maximum is also in all things. And because it is absolute, it is, actually, every possible being; it contracts nothing from things, all of which [derive] from it.'

Nicholas saw Brahman through his mind. He called it "Learned Ignorance". Oh how different things would be if history followed his thoughts. But we shan't cry over what did not come to pass.

As Nicholas states, the idea of absolute infinity is indistinguishable from the idea of one. That is the key to the whole thing. Thus we come to the main insight of this book:

To say "reflect the self in the not-self" is to say "infinity tries to copy itself". It makes no sense. It is logical nonsense as if to say "1 = 2", or "true = false", or "I am lying" or any other such logical travesty.

If you are a Clever Inquirer you will counter: it is the infinite; it contains all things, even all contradictions. Why cannot the infinite copy of itself?

And I simply say to you: look thus about you at the result.

The result is seemingly infinite spirals chasing their tails but never catching them: potential infinities that spin and spin, trying to be real infinities; processes that seem to be able to go on forever, but cannot. Said differently, if the myriad potential infinities were really absolute infinities, all would be complete and at peace. There would be no movement.

Addressing the Clever Inquirer

There can only be one absolute infinity. One and absolute infinity are self-identical. Not even equal, because equal implies the equivalence of two different forms. This is not the relationship between one and absolute infinity. There is no relationship between one and absolute infinity because they are the same thing, self-identical.

It is like saying I have a relationship with myself. This makes no sense because "I" and "myself" are the same thing. I can have relationships with my thoughts, my feelings, my body. But I cannot have a relationship to myself. The very concept is nothing more than a tautology between words that mean exactly the same thing, like saying "I am me".

Thus, the answer to the Clever Inquirer is that Maya, the presence of potential infinities within absolute infinity, is not a contradiction. Yes, infinity can contain all contradictions. But Maya is not a contradiction.

Consider the issue from another point of view. Take the following two statements:

1. Absolute infinity is one.
2. Absolute infinity copies itself to make many absolute infinities.

These are not two contradictory cases. They are different definitions of the term "absolute infinity". They are mutually exclusive definitions. If you accept definition 1, then definition 2 is ruled out, and vice versa.

Definition 2 is self-contradictory. If you can have two absolute infinities, then neither is absolute. This is the road gone down by Cantor. It is the set of all sets containing itself. It is grounded in the illusory notion that one can manifest absolute infinity in some form. But this is impossible because form implies boundaries. Absolute infinity is unbounded in any possible sense you can imagine.

Thus, Nicholas of Cusa hit the nail on the head at the very beginning of the modern era when he realized that there is only one instance of absolute infinity.

The crux of the argument however does not turn on the logic, but appeals to *experience*. Look about you at the world. We only observe potential infinities in nature. We only produce potential infinities in our minds. All nature is striving and incomplete. In all our experience as humans, we have not one manifest example of absolute infinity. It is only an ideal. Inductively it is always possible that tomorrow an absolute infinity will manifest and engulf everything. But this line of thought is a joke. Absolute infinity already engulfs everything. Hindus call it Brahman. I would say "absolute infinity is starring us in the face", but it is because absolute infinity *is* that we can stare at anything.

Therefore, let's summarize the philosophical argument under consideration, which, recall is Krishnananda's not mine. I am merely the messenger, and I grant that I am by no means the best possible messenger. Be that as it may, the main point is the following:

Krishnananda's idea that experience derives from the drisimatrah projecting its being into the not-self can be reframed as the absolute trying to express itself as the relative. The act seems successful but really is not. It seems successful insofar as it generates our existence. But our existence is not being, it is becoming, an ever-striving to be whole and complete, which eludes us on every level. Infinity does not copy itself but generates potential infinities of which we are instances.

It is the ancient concept of the Many in the One. It is an idea that echoes Leibniz' idea of monads and God. The monads are little potential infinities; God is absolute infinity. Unlike Leibniz, the yogic view does not see a perfect pre-established harmony amongst the monads, only illusion and senseless chaos. Cosmos is a momentary illusion of Maya.

Voltaire's mockery of Leibniz' "best of all possible worlds" is thus reflected in the yogic view. However, yoga does not mock. It takes the whole matter very seriously and seeks to resolve the issue.

The means by which the Absolute expresses the Relative, the One becomes the Many, are beyond our ken. The very inscrutability of Maya, the inability to express it intellectually, to resolve it in any practical way within our experience of becoming, this is the meaning of avidya: An ignorance of such epic cosmic proportions that it creates … well … everything.

Chasing Mirages

The previous section is wholly abstract. To most people it will seem as so much mumbo jumbo. That is okay. It is stated for those who wish to hear it. It is a perspective that may be helpful to some people.

There are more down-to-Earth implications of projecting the being of our self on to the patterns of flux which appear in our minds. This projection process is the essence of desire, of want and longing. We desire a thing because we think it will bring us some type of satisfaction, fulfillment, or completion.

The seemingly simple and obvious facts of life—that we desire, and long for, and wish, and need—is the consequence in our immediate experience of the inscrutable process of Maya twisting the drisimatrah into potential infinities.

As Alan Watts quote indicated in Chapter 2, to understand where our desires and needs come from implies the entire under-surface of the mind:

> "…[The] simultaneous presence of all the levels of man's history, as of all the stages of life before man. Every step in the game becomes as clear as the rings in a severed tree. But this is an ascending hierarchy of maneuvers, of stratagems capping stratagems, all symbolized in the overlays of refinement beneath which the original howl is still sounding."

In Hindu thought, these echoes of previous experience are called

samskaras. Samskaras are memories of all past acts at all levels of manifestation. The term implies the inter-relatedness of all things, which, as mentioned previously, is the nature of the Relative.

Going to this level of depth will get us back to the mumbo jumbo in the previous section. We don't want to do that. What needs to be pointed out is that yoga shows a direct connection between the abstractions considered in the previous section, and our day-in-day-out desires and needs. Having said this, however, let us confine the discussion to the surface mind.

What is desire? It is the expression of incompleteness. It is the recognition of an emptiness that is then sought to be filled. The emptiness experienced, the hole, is the samskara, the tendency towards a particular something. The act of trying to fill the hole is karma, action.

Why is there a perception of a hole, an emptiness, in our being? Obviously there are as many specific reasons as there are specific desires. But ultimately desire in general exists because we are manifestations of incomplete potential infinities.

Desire is an awareness of discomfort. Discomfort of the body when the body experiences needs like hunger or being tired. Discomfort of the emotions when there is anger, threat, or when love goes unfulfilled. Discomfort of the mind when goals are not accomplished, when perceived needs are present (as in "I need an iPhone"). Why is desire a state of discomfort?

According to Krishnananda, desire creates discomfort because of the contorted posture the drisimatrah takes in pretending the not-self has being. The very act of projecting being on what is not the Self is the source of the discomfort. By projecting its being on something else, the drisimatrah is effectively trying to cut its indivisible unity into pieces. Again, it is an impossible act. The result is a type of tension that produces awareness of discomfort. Ultimately, the discomfort traces back to the tension of innumerable Eenie-Weenies, each trying to fuse into itself, but unable to do so.

The discomfort of desire plays out in innumerable ways. We want something. Whatever it is: good food, status, comfort, safety, love, sex, whatever. We can only achieve these desires by finding something outside of our self, which then becomes the object of our desire. We believe that in obtaining the object of our desire that we will find fulfillment.

There are two general cases. We will either be successful or we will not be successful in acquiring the desired object. If unsuccessful, then my point is made: we are left wanting, incomplete.

In the case we are successful and acquire the desired object, the case is more complicated.

We acquire the desired object and there is momentary satisfaction. This is the pleasure spoken of earlier, that taste of honey that cuts your tongue and really doesn't taste so sweet.

According to Krishnananda, the satisfaction obtained from acquiring an object of desire is really just the momentary release from the tension that generated the desire in the first place. The release from the tension is what we call "pleasure". But it is really not pleasure, just the momentary release from a discomforting state. It is less painful, and therefore perceived as "pleasure".

Then another factor comes into play. Since the object of desire is not a part of the self, it will eventually separate from the experience of the self. Then there will be longing and the need to repeat the experience. The longing adds to the existing tension. The desire will again exert its discomfort and the cycle begins anew.

The next time the desired object is acquired there is memory of the pain of having previously lost it. This makes the release of tension less for each additional time the object is acquired. The pleasure is now not even sweet, but bitter-sweet.

Then, the desire repeats over some number of cycles. Eventually we lose interest. The original cause of the desire is exhausted. Then new desires arise to replace the previous ones, and the whole cycle starts again.

So, it is seen that desire in general cannot be satisfied by any object fully and completely for all time. Instead, desire is a vicious cycle.

Some concrete examples we all know only too well are the following. Maybe that job you wanted did not bring the status and satisfaction you had hoped for. The man or woman you married was not the "soul mate" you thought. Things change. The job that was exciting starts to become routine, perhaps boring, maybe even depressing: the same with your romantic interests, or your hobbies. You adapt. Maybe you adapt by changing your attitude to the new circumstances. Maybe you discard the old circumstances and seek again to find your fulfillment in a new situation: a new job, a new marriage, a new hobby, whatever it may be.

To summarize:

Case 1: We do not get what we want and therefore carry around the constant tension of the unfulfilled desire. Not that I will dwell on it here, but the unfulfilled tensions can amplify over time and may turn into a variety of psychic pathologies, or they may not and you will be just fine.

Case 2: If we do get what we want, the situation transforms before our very eyes. We started with A, which turned to B, which has led us to a new circumstance, C.

And then, to put the icing on the cake: You die.

Be honest. Did you really get what you wanted?

So what's the point? What's the point of all this striving? This is why people say stupid stuff like: "oh do it for the children!" People seek some excuse, any excuse whatsoever, to justify this absurd and seemingly meaningless merry-go-round of chasing after mirages. The kids could care less. They are immune to your consultations and are busy living their own lives.

So, we do not need to understand any of this at fancy intellectual levels. All you need to do is look at your own life, your own experience. Look at how you seek happiness in the things outside of you. See yourself projecting your desires on to objects outside of you. You are seeing the Self in the Not-Self, and thereby making experience.

This is the fundamental mistake that lies at the very core of our experiences in this world of seeming. You grab at the other, trying to make it a part of you, trying to find fulfillment and completion. But it is not part of you. It is its own thing, with its own existence. You were wrong in the first place to chase after what you are not. If you happen to grab it for some amount of time, it will eventually slip away from you, like sand slipping between your fingers.

Then you will die.

So there. Let's just stare directly into this inscrutable condition of our experience. See that we can never understand it. Know that we will die and not understand it. Stare at it in all honesty and tremble in the face of our terrible condition. Stare at avidya.

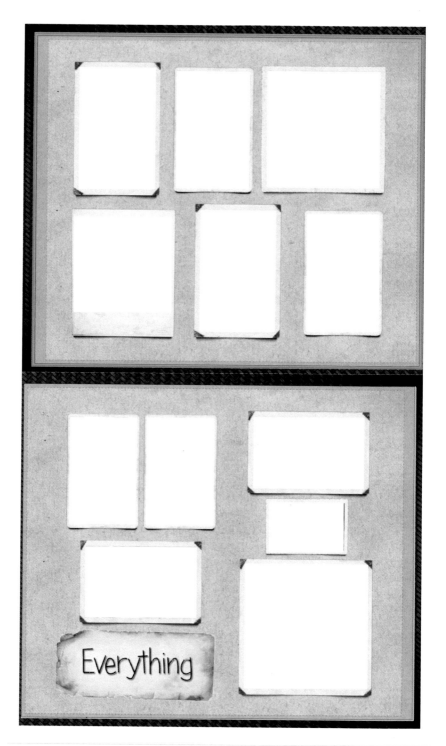

⅋ Experience: Everything

What's The Point?

What is the point of the ghastly picture we have painted about our experience? First let's review and summarize that picture.

The general picture can be stated briefly and succinctly: our experience consists in chasing after mirages, then we die.

How do we know something is a mirage and not real? A mirage disappears when inspected beyond its superficial appearance. We find nothing there at all. Similarly, for all our struggles, striving, desires, and efforts to get what we want, we just die. The whole thing evaporates like a mirage.

For those who can appreciate things beyond the physical: whether there is life after death doesn't matter per se. Even if something in us—a soul or whatever—survives physical death, for most souls, they just keep chasing mirages, but now in the inner realms. The impact our departed soul has on the physical world becomes nil, in spite of everything we were and everything we did when alive. The world goes on without us and sooner or later we are completely forgotten.

Just look at Alexander the Great's empire. Or recall the names and deeds, the glories of all those extremely important Roman Emperors on whose whim whole societies rose and fell. Let's converse of fond memories of our favorite rulers of the Umayyad Caliphate. Tell me your fond reminiscences of the Achaemenid Empire.

Oh wait...you can't.

So, while it seems today that there are great men and women whose legacy will last forever, time will eventually consign even them to oblivion, just like all those other extremely important people in history, you know; the ones we can't remember or never heard of.

Why do we chase after mirages? Because we feel a sense of incompleteness that we perceive as desires of all stripes, and we strive our whole life to fill those holes. Why the sense of incompleteness? Well, I've offered the abstractions of Eenie-Weenies, potential infinities, and Maya.

Whether these ideas are accurate reflections of experience or not also does not matter. When stripped of all the abstractions, it's really not very difficult to understand. We chase after mirages, pretty much to no end, because, in the end, we just die and, sooner or later, the sands of time erase us from the face of eternity.

So we come back to the question: what's the point of all this? Here we are, after all. Why are we in this condition? What can you do with this picture of experience? What good is it? Can it serve a productive end?

The short answer is: yes.

Ready...Set...Go!

If we take any other approach to our experience, particularly those that are optimistic (like God created it, or there is evolution and progress, or we must sacrifice for the children, etc. etc. ad nauseam), there are two consequences:

[1] We are lying to ourselves because life pretty much does suck for all the reasons stated above.
[2] By painting some kind of rosy picture of things, we keep our selves confined to this crappy condition by construing it to be something that it is not. Or said slightly different: truth does not come to the deluded.

The short of it is: you only screw yourself by lying to or deluding yourself about the condition in which you find yourself.

It comes back to the conclusion of Chapter 4: when we see through the mirages of our experience, whether on the surface of the mind, or those underneath it, it should trigger off a new desire: to make it stop.

When our condition is understood, the only sane response is to want to make this whole nightmare disappear, to get off this insane merry-go-round of mirage chasing, to be rid of this condition once and for all.

This then leads us to ask: what else is there? What else can there be?

That is the main reason for staring right into the heart of the darkness of our experience, to stare at the incomprehensible insanity of it all. Because then we stimulate in ourselves the desire to go beyond this condition. We begin to wonder if there is a way off this insane merry-go-round. We begin to wonder if it is possible to be in a state where there are no mirages.

It is only then we are ready for yoga. The first aphorism of Patanjali's *Yoga Sutras* is:

१.   अथ योगानुशासनम् ।

Atha Yogānuśāsanam.

"Now, an exposition of Yoga (is to be made)."

To a Western person, this short sentence just says "now yoga will be discussed". But to one who really understands the overall situation of our experience, a whole dissertation can be written on just this first aphorism. Here is a 48 minute talk[5] on just this first aphorism. The previous chapters of this book are a commentary on just this first aphorism of the *Yoga Sutras*.

Unlike the Western use of language where a sentence is what it is at face value, an aphorism in a sutra is like a seed, a genetic code, and it can be unwound and unfolded into the most complex of thoughts. Each

---

[5] http://www.youtube.com/watch?v=HMfGELH-ZTw

aphorism is like a formula that contains layer after layer of meaning.

The Hindus figured out how to do this many thousands of years ago when books were rare, people were generally illiterate, information was orally transmitted, and means were required to convey a lot of information succinctly. We Westerns have no appreciation of the sutra method. Go study the commentaries of the *Yoga Sutras* and see this method in action. It is impressive.

I will now unwind just a small bit of meaning in the phrase "Atha Yoganusasanam".

This phrase does not just mean "now yoga will be discussed". It also means something to the effect:

> "You have come to me seeking yoga. This implies you have now experienced life sufficiently to understand the necessity to study yoga. You have seen through life's illusions, allures, sorrows, and limitations, and are now ready to take the next step. Your soul has matured so that you are no longer entranced by the mirages and enchantments of life and Maya. You are world-weary and done playing on the merry-go-round of the Maya. You sense there is something greater. Listen to me then, for I am now to elaborate on those things that are beyond the illusions and sorrows and mirages of your existence. I will now speak of how to resolve this seemingly unsolvable condition in which you find yourself."

Something along these lines is what the phrase "Atha Yoganusasanam" really means. It means *you are ready* to learn yoga.

The Light At The End Of The Tunnel

The second sutra then tells what you will learn. It is very famous. It is the text-book definition of yoga: ***yogah chitta vritti nirodhah***.

२. योगश्चित्तवृत्तिनिरोध : ।

Yogaś citta-vṛtti-nirodhaḥ.

"Yoga is the inhibition of the modifications of the mind."

Again, a whole dissertation could be written on the meaning of this short phrase. Many such have indeed been written by some of the greatest minds of Humanity and also by lesser minds too, like me.

This book, *Experience*, is also a dissertation on the nature of the **vrittis**, the patterns found in the mind. We have described these as the mirages of incompleteness, as potential infinities that make up the patterns of nature perceived by the physical brain and by consciousness at all of its deeper levels. The infinity of mirages of ever-ascending greatness are the vrittis. Taken as a whole, Hindus call this pattern: **Prakriti** or **Shakti**.

At this point, we could enter into a dissertation on the methods of yoga, but we will not do that here. For the interested Reader, I have written a summary of the overall methods of yoga[6], and a summary of the more advanced methods[7]. These summaries are meant as introductory texts. For those who care to learn more, an almost unlimited supply of additional material, spanning many thousands of years, is available if one just looks for it.

Where we wish to go now is to wrap up *Experience* with a discussion of three aspects of our experience that lead directly to the methods of yoga. These thoughts fall somewhere between aphorisms 1 and 2 of the *Yoga Sutras*. By this I mean that, once we start to see through the mirages of relative existence and begin to question if there is anything beyond this condition (as implied by aphorism 1), then we need to find "anchors" or "hooks" in our present experience that serve as seeds that naturally lead us into the yogic practices, called **abhyasa**, that are implied by aphorism 2.

The three aspects of our present condition that naturally bridge us into yoga are:

[1] The capacity for vairagya
[2] Drisimatrah
[3] That we naturally and spontaneously move between different states of consciousness

---

[6]http://dondeg.wordpress.com/2014/05/02/what-is-science-part-6-the-methods-of-yoga/
[7] http://dondeg.wordpress.com/2014/05/28/patanjalis-ten-types-of-samadhi/

Learning Vairagya

The critical, dispassionate, philosophical analyses that allow one to begin to see through the mirages and allurements of this life are a "hook" into the methods of yoga.

Such thinking will not cease if you begin practicing yoga, but will grow into understand that dwarfs your present comprehension. Such critical thinking is the basis of **vairagya**, dispassion; the attitude required to be successful at yoga. Yoga, ultimately, is wholly impersonal. Dispassionate insight is gained by using the intellect in ways unknown to Western methods.

The so-called "objectivity" of modern science is but a pale and feeble reflection of the dispassionate impersonality of vairagya. Modern science is not objective in any deep sense because it is blind to the role of the Self, the mind, and the senses when seeking to understand the nature of the world, as I discussed in *What Is Science?* However, the *ideal* of objectivity in science gives us an initial seed idea to begin to understand the meaning of vairagya.

Experience has been an exercise in vairagya, by attempting a dispassionate dissection of our condition. It doesn't even matter if there are mistakes in what has been said throughout. What is important is my honest intent to find the truth. If the desire to find the truth is sincere, my mistakes will be corrected with time. What is important is the sincere attempt to be critical, thoughtful, truthful, and intellectual. Together these serve as the foundation of vairagya.

Vairagya is critical from the beginning to the end of yoga. It dispels the mirages of life at the beginning, lets us see that invitations from celestial beings are something to worry about in the middle, and allows the final jump at the end (this last one symbolized by the Temptation of Christ).

The External is Not Eternal

We discussed drisimatrah in Chapter 7. The fact that we are conscious and self-aware is the basis of all yogic practices. Like vairagya, it is there from the beginning to the end of yoga, and continues beyond.

Krishnananda's statement "The External is Not Eternal" provides the central insight for understanding how drisimatrah can extricate us from the illusory state in which we find ourselves.

What does it mean to extricate ourselves from the condition of incompleteness and relativity? Where do we go?
The answer to these questions is given in the 3rd aphorism of the *Yoga Sutras*:

३. तदा द्रष्टुः स्वरूपेऽवस्थानम् ।

Tadā draṣṭuḥ svarūpe 'vasthānam.

"Then the Seer is established in its true nature"

This aphorism describes the end result of yoga, the final outcome. Yoga means "the joining" and the joining is here specified: The Seer joins with itself. We can state this alternatively in terms we have used throughout:

[1] Drisimatrah quits projecting its being seemingly outside of itself. When this process of projecting ends, drisimatrah dissolves back into itself.

[2] Absolute infinity quits trying to twist itself into relative potential infinities. The potential infinities evaporate and absolute infinity is calm in itself.

[3] We no longer experience the emptiness of desire for worldly things. We find our Self to be whole and complete and there is no longer a need for attachment to worldly objects.

What all these describe is the state that is alternative to the merry-go-round of relative being. They describe where one "goes", so to speak, after escaping from the looney bin of becoming. They are three different ways to say the same thing; four if you include Patanjali's way of saying it. What they all say is that consciousness dissolves back into itself and becomes free from the vrittis and samskaras that put consciousness, drisimatrah, into the state of relative becoming and incompleteness.

Tangent for a Summary

At this point I must pause and marvel at the structure of only the first 3 out of 196 of the aphorisms of *The Yoga Sutras*. In just 3 sentences, an immense amount of information is condensed into the most perfect of logical structures:

"You are now ready to learn yoga"
"Yoga is the removing of the patterns from consciousness"
"When this is accomplished, consciousness will be alone within itself"

It is a marvel of language.

It is unfortunate that post-modern deconstruction of language has developed along the hostile lines it has. It seems if one wished to focus on studying language, then a deep study of the Sutra method would serve well the future intellectual development of Humanity. The sutra method has obviously been productive for the Hindus, but it needs to be widely disseminated to all of Humanity.

Ok, back to the regularly scheduled program...

Paranga Cetana and Pratyak Cetana

By studying aphorism 3, we can begin to appreciate more deeply Krishnananda's statement, "The External is Not Eternal". The idea is that the condition of **drashtuh svarupe avasthanam**, consciousness absorbed in itself, is the condition of pure internality. There is no "external", no "object" in this condition. If we perceive anything as external to our self, then we know we are in the realm of the relative. But how can we *not* perceive something as external to ourselves?

There is a crucial set of ideas in yoga that explain what this means and also provide the bridge between our "normal" relative state and the methods and goal of yoga. I discussed these ideas here[8] and here[9] and now expand on the importance of these concepts.

---

[8] http://dondeg.wordpress.com/2013/11/16/mobius-strips-the-bindu-and-moving-amongst-the-planes/
[9] http://dondeg.wordpress.com/2014/05/28/patanjalis-ten-types-of-samadhi/

The key concepts are **paranga cetana** and **pratyak cetana**. Paranga cetana is consciousness directed outwardly. Pratyak cetana is consciousness directed back on itself.

Paranga and pratyak cetana are the most important concepts I have ever learned. The realization that consciousness can be either outward-directed or inward-directed explains more than any other concepts I know. These ideas are not intellectual abstractions, but like drisimatrah, refer to something we constantly experience.

Recall Taimni's diagram from *The Science of Yoga*:

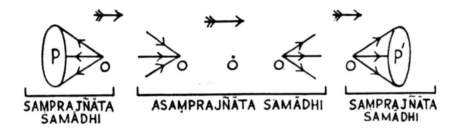

In this diagram, 'o' is the center of consciousness, the thing Alan Watts discusses in Chapter 2. This is the center from which our subjective awareness, our sense of being, drisimatrah, projects forth. On the left and right (labeled "samprajnata samadhi"), the arrows go from "o" to "P". In this diagram, "P" stands for "pratyaya", which is the technical term in yoga for "object of meditation". However, "P" can generally stand for "perception", the perception of anything whatsoever that is external to the perceiving self. When the arrow goes from the center of consciousness towards *anything*, this is the condition of paranga cetana.

Everything discussed in *Experience* refers to the condition of paranga cetana. It is the only condition of consciousness most people know, even those who have experienced altered states of consciousness. The concept applies to people trapped in the surface mind as well as to people who have experienced the inner realms under the surface mind. The common feature of the surface mind and the realms underneath is that **they all appear to be outside of the perceiving consciousness**. They all are states of paranga cetana.

If something appears to be outside of the perceiving consciousness, whether that something is the physical world, the worlds of mind, or deeper "spiritual" worlds, it is still a condition of paranga cetana.

Paranga cetana is the condition of the Relative, of incomplete infinities, of unfulfilled promises, of mirages. Paranga cetana is the condition where drisimatrah projects its being onto objects seemingly outside itself, generating the tension and discomfort we experience as desire, want, and longing.

Then The Seer Abides In Itself

Now, if even the perception of the inner realms below the surface of the mind is paranga cetana, then what is pratyak cetana? Notice in the diagram, there is only one side "o" and the arrows point at "o". What does this mean?

As I indicated above, it means, that, in some sense, consciousness projects back into or onto itself, or dissolves back into itself. This seems abstract as an intellectual matter. But it is something we experience every time we transition from being awake to going to sleep. The transformation of our physically-aware consciousness to any other state of consciousness involves the process Taimni depicts above. It is thus his diagram explains the mechanism of how consciousness moves amongst its many states.

When we pass through the center of consciousness, it seems instantaneous. It seems to take no time to transition from one state of consciousness to another[10]. This is because consciousness dissolves into itself, and consciousness is eternal and transcends time. We are, for that

---

[10] An abrupt and instantaneous transition is called a "bifurcation" in modern dynamics. The transition from waking to sleep is accompanied by graded changes in neurotransmitters and regional brain activation patterns. However, at specific thresholds of these, the net result is an abrupt bifurcation in the global brain state, correlating with the transition in the state of consciousness. Bifurcations occur instantly with respect to control parameters. In some way, which I have yet to explore, the abrupt physical transition must be related to the ancient yogic insight conveyed in aphorism 4.33 of the *Yoga Sutras*. In short, what I am saying in the text does not contradict our current scientific understanding of the brain-mind system.

instant, pure drisimatrah and everything that seems to exist evaporates in that instant. Time is replaced by eternity. Space is replaced by absolute infinity.

I know the previous paragraph sounds kooky. But this is exactly how it is described in the *Yoga Sutras*. I have written about this here[11] and will not repeat myself. The short of it is: consciousness slips *between* the moments of time and thereby moves out of time and into eternity. This is aphorism 4.33 for any Readers who wish to follow up on the issue.

The seemingly momentary transition of pratyak cetana, when "The Seer Abides In Itself" seems devoid of any content at all, leading one to think it is a state absent of drisimatrah. But it is the exact opposite. It is pure being within itself.

The fact that pratyak cetana exists is why the Eastern approach will not just result in another empty promise. Pratyak cetana is the escape hatch off the Merry-go-round we call experience.

Of course, the natural and spontaneous momentary experience we have of pratyak cetana when we fall asleep is but a seed experience. It is built naturally into the body but must be cultivated to fuller expression. The whole study and practice of yoga is the study and practice of expanding this momentary experience to encompass all of our being, which is all being, which is The Absolute.

The Beginning

As people, as real human beings, we are seemingly trapped in the infinite echoes and mirages of relative existence. This is the condition of outwardly directed consciousness, of paranga cetana. We have reviewed its constitution. We call it "infinity" and imagine it contains everything within it. But it is all illusion. It is but appearances of things that promise to be but never are.

The whole phantasmagoria of becoming proves to us that there is something above, behind, and beyond it. Again, a mirror cannot show a

---

[11] http://dondeg.wordpress.com/2014/05/28/patanjalis-ten-types-of-samadhi/

reflection unless it reflects something. By analogy, the seeming of our relative becoming—the state of our experience as unending incompleteness—is like a mirror image. There must be something reflecting in the mirror. There is: Drisimatrah, our very consciousness itself. It sees itself reflected in the patterns of relative, transient fluxes of becoming and misconstrues its own nature.

Yoga shows the escape hatch; how drisimatrah can ever so slowly, step by step, layer by layer, dissociate itself from its reflections. As this process continues, drisimatrah recedes back into itself free of the reflections. The Seer Abides in its own Nature.

To summarize: that all of this is not fantasy is seen in two basic facts of our experience. First, we are. Within our "Is-ness" we are aware, and we are aware that we are aware. That is the signature of drisimatrah, of truth, of absolute infinity. Second is the fact of pratyak cetana. Even those trapped in the surface mind still shift amongst waking, dreams and dreamlessness. For those who have had the curtain pulled back, they know there are other states of consciousness, which are also mirages of relative existence. In either case, the second signature is there—pratyak cetana—the ability of consciousness to be within itself.

Drisimatrah is obvious. You only need to imagine your self-awareness free of all distractions, vrittis, desires, and movements to imagine drisimatrah.

Pratyak cetana appears as oblivion to those who have not yet learned to see that the ever-becoming and mirages of outwardly directed consciousness, of paranga cetana, is the true oblivion.

The Absolute Truth of pratyak cetana appears as but nothingness to those who have not learned to see its eternal, unspeakable brilliance, its infinite splendor.

> "...when any of them is liberated...the glare will distress him, and he will be unable to see the realities of which in his former state he had seen the shadows..."

> - *Plato Book VII, The Republic*

Seattle Post-Intelligencer

AMERICA FIRST — CHARACTER · QUALITY · ACCURACY · ENTERPRISE

Post-Intelligencer
Want Ads
Lead the Way
to Quick Action
Results
Call MAin 2000

VOL. CXIII, NO. 27   SEATTLE, WEDNESDAY, SEPTEMBER 29, 1937   THIRTY-TWO PAGES   DAILY 5c, SUNDAY 10c

TRAPPED IN MIRAGES

Should I Stay or Should I go Now??

Was Plato a Total Freak? Was he Crazy As a Loon?

What to do in the meantime? How to Keep busy..

Chief Executive at Bonneville Pledges Widest Possible Use Of Tremendous Resource

Full page of pictures of President Roosevelt's Northwest tour on Back Page.

By John Boettiger

BONNEVILLE DAM, Ore., Sept. 28.— The great power resources of the Columbia River will be made available to all the people of the region at lowest possible rates, it was pledged today by President Roosevelt in his address at Bonneville Dam.

Huge Crowd Lines Parade Route; Executive Smiles at Packed Throngs on 2nd Ave.

Experience: Epilogue

It seems appropriate to end *Experience* addressing the issue: okay, so what now? What is one supposed to do? If all of life is just chasing after mirages, and the so-called "escape hatch" involves some bizarre process of having consciousness rest within itself, what is one supposed to do in the meantime?

To address this concern we start by realizing that knowledge of yoga, of *real* yoga, not this silly Westernized hatha yoga stuff, puts one at a decision branch point. When the yogic view of our experience is explained there are two general reactions: (1) either you reject it, for whatever reasons, or (2) you accept it, for whatever reasons. Your reaction naturally puts you into one of two camps: the Rejecters and Accepters.

Rejecters

Many people are inclined to reject the yogic assessment of life. They find it hard to believe that all the things they love and like, things that are dear to their heart, are mirages, or that there is anything wrong with the way things are. Never mind the fact that your death is imminent. For most people, this doesn't matter. They ignore the inevitability of death and cling to whatever it is that life has given them, right up to the very end. They are not going to trade the few drops of honey life has given them for some nebulous possibility of "absolute infinity", for some completely unimaginable state that Plato described in his Allegory of the Cave.

Mainly this is the reaction of those confined to the surface mind. However, it can also happen to those people who have seen below the surface mind, but who may be ensnarled in some illusion of the Intermediate Zone.

This is a perfectly legitimate reaction. It means that one is not ready to embark on the grand adventure of yoga. One is still bound to the attachments of life and is not ready to let go yet. The state is similar to an unripe fruit: a green banana or green orange. The unripe orange will not fall off the tree. It won't fall off until it is ripened sufficiently, then it automatically falls off.

Further, from the Hindu point of view, death of the physical body is not the end anyway. The person dies, exists for a while on the inner planes doing whatever, and then reincarnates again based on their samskaras. This is samsāra, the wheel of birth, death and rebirth. This is the merry-go-round. You—you soul—simply recycles on the wheel until you are ready to move on.

This is a great beauty of the situation we find ourselves in, or you could say, the beauty of God's wisdom: Since it is all mirages, it doesn't matter if we do or do not chase them. No fundamental harm is done in either case. Unlike the evolution viewpoint that ultimately reduces to randomness, and hence meaninglessness (or some nebulous, emotion-laden view of "creativity"), the yogic view is more like religious views that see God's plan in all events and occurrences. Every soul has their moment in the scheme of things when the time comes to let go of the mirages and make the move to deeper realities.

Yoga is very clear about this: if it is not your time to fall off the tree, we don't want you. Go back to your illusions until you are serious about dispelling them.

Accepters

For those who accept the yogic assessment of life, who are ready to embark on getting off the merry-go-round, it becomes a serious issue what to do in the meantime. Just because we realize that we are in this dumb condition of chasing mirages doesn't mean it's going to just magically stop once we realize what is going on.

The whole process of wanting to get off the merry-go-round of samsara is kind of like a car skidding to a stop. Just because you have applied the brake doesn't mean the car instantly stops. It doesn't. A moving car has momentum, and that momentum must be used up before the car stops. It is the same with samsara. Over many lives, you've built up a type of karmic momentum. All of this "energy" needs to be used up before you can exit the merry-go-round. Again, this "energy" comes in the form of the samskaras, the tendencies to desire and want specific things.

If you read the *Yoga Sutras*, probably close to half of it is dedicated to explaining about the samskaras (including the Kleshas) and what exercises you need to do to exhaust them. It is all very involved. I am thinking about writing a multi-part essay on the topic of samskaras and karmas because these ideas are still not adequately understood in the West.

But the main point is, the momentum of the samskaras just keeps on going, whether you are aware of your condition in the mirages or not. So the issue becomes one of intentionally starting to exhaust your existing samskaras.

Starting Points

There are two general ways people come to realize the truth of the yogic position: (1) intellectually, or (2) experientially via some degree of mystical experience. These may occur in the same person, or it may be that just one or the other has occurred.

The intellectual way is the weakest because it is only a pattern of thoughts that serves as a bulwark against the momentum of the samskaras. At any moment, new ideas could come in and disrupt the apple cart and toss one right back into the realm of mirages. In addition, unless the intellectual understanding is grounded in deep insight, one's emotions and sentimentality can also disrupt the intellectual understanding. If such emotional disruption occurs, it means the intellectual understanding was indeed very weak from the start. There must be constant reinforcement of the intellectual understanding to sustain this path.

If one gets a glimpse into the pure conscious condition (drisimatrah), this is stronger because one has actually *experienced*, to a greater or lesser degree, what is at stake. This can be less subject to doubt than mere intellectual arguments and strengthen the mind towards yogic practices. On the other hand, one must not get caught in Intermediate Zone illusions. The mystical path is plagued by the ambiguity of knowing if it was a real experience or not. One can determine the value of the mystical experience by whether it triggers in you only a minor, a medium, or a strong desire to tread the path of yoga.

In either case, one must realize that just knowing that the path of yoga exists is no guarantee one will enter this path, let alone stay on it. The end of *Experience* describes the *beginning* of yoga, and like any nascent situation (newly planted seed, newly implanted embryo), there are many ways for the whole process to naturally abort. [For those who do not know, in humans, over 2/3rds of acts of conception end up aborting, just naturally].

There are no rosy pictures of yoga. Once one chooses this path, the first step is to stick to it, which is difficult. There is nothing fun per se about the ensuing stages. The process of deconstructing one's own ego (asmita), of peeling back layer after layer of illusion, is arduous. One only sticks with it because at some level or another, one knows there is no alternative, because one really is serious about getting off the wheel of illusions.

In the Meantime

So, for the Accepters, what to do in the meantime? Swami J has a most helpful discussion[12] of the options open to newcomers to the path of yoga. There is no simple answer, no "one size fits all" approach. Everyone is different and will respond and act according to their innate capabilities.

I will close out here giving my own personal approach on the matter,

---

[12] http://www.swamij.com/yoga-sutras-11922.htm

which has become increasingly clear as I learn more about what the *Yoga Sutras* teaches. An important insight from Chapter 18 of Krishnananda's *Study and Practice of Yoga* is:

> "Self-control is the introduction of some element of the nature of Truth into the perceptions of the mind, and would be the first step of control of the modifications of the mind-stuff. **We cannot control the mind by the force of will**. Every stage in the practice of yoga is really a positive step in the sense that there is a healthy growth into new stages of Reality, rather than merely a withdrawal from unreality." [Emphasis mine]

He elaborates this idea in great detail, which I only summarize here. The idea that one must meditate *is* the center of yogic practice. But it is also an advanced stage of yoga. There are preliminaries that must be in place for meditation to be successful. You would not expect a kindergarten student to do sixth grade work, nor a sixth grader to do college work. It is much the same with yoga.

Krishnananda points out something very important about yogah chitta vritti nirodhah. It is not just stopping the flow of thoughts by sheer will power. It is the replacing of perceptions of illusion with perceptions of truth. It is by coming to understand the truth of our condition that the mind, eventually, and of its own accord, will slow to a silence. Nothing is forced. Everything occurs naturally and gradually. This is a key insight.

At the earliest stages, one must not give in to the temptation to try to force the mind into any condition. Krishnananda spends many pages talking about how tricky the mind is. It is like a wild beast that must be tamed, not by force, but by truth.

This comes back to the 8 steps of practice of Patanjali's *Yoga Sutras*: yama, niyama, asanas, pranayama, pratyahara, dharana, dhyana and samadhi. The first steps are yama and niyama.

Because there is so much focus on meditation, which begins at asanas, and because Western civilization has a mass case of Attention Deficit Disorder and wants instant results, people jump right into the meditation aspect and ignore the first two stages of yama and niyama.

But this will not work. Yama and niyama are the foundation of vairagya, the dispassionate attitude required for success in yoga. What I am getting to can be said quite simply: for a newbie in yoga, one can spend the rest of this life just practicing yama and niyama, and it would be very productive.

There is another important word in yoga that has not yet made it into the *Experience* essay that is intimately linked to vairagya, and that is **viveka**. Viveka means "discrimination" as in the ability to discriminate true from false. Let us consider Taimni's words from the *Science of Yoga*:

> "Real Vairagya is not characterized by a violent struggle with our desires. It comes naturally and in its most effective form by the exercise of our discriminative faculty which is called Viveka. Glamour plays a very great part in producing Raga or attachment and even ordinary intellectual analysis combined with reason and commonsense, can free us from many unreasonable habits and attachments. But the real weapon to be used in acquiring true Vairagya is the more penetrating light of Buddhi which expresses itself as Viveka. As our bodies are purified and our mind becomes free from the cruder desires this light shines with increasing brightness and destroys our attachments by exposing the illusions which underlie them. In fact Viveka and Vairagya may be considered as two aspects of the same process of dissipation of illusion through the exercise of discrimination on the one hand and renunciation on the other."

Again we see the emphasis on a natural dissipation, not a forced suppression, of the movements of the mind.

So, what to do in the meantime? If one can accept the logic of yoga (and of course other traditions like Buddhism) that strips the veneer of rosiness and glamor from life, then what to do in the meantime is strengthen one's mind.

This strengthening comes in the form of ignoring the glamours and illusions, of seeing through them and naturally losing interest in them. At the same time, one studies and learns how things work. How the mind works, with all its tricks and glamours; how the mind itself is like a fun house of mirrors. One comes to focus more and more, not on things of this world, but on spiritual truths, the scriptures of the world. Whichever ones appeal to us the most, whether Christian, Hindu, Buddhist, etc.; they all say effectively the same thing.

One begins to slowly divert their desires away from the illusory mirages of the external world, and towards God, consciousness, infinity, whatever you want to call it. In yoga, this is called "bhakti" or a devotional attitude. But there is nothing goofy or emotional about it. It arises from a serious and sober understanding of things.

A rearrangement of the mind is required. To this point, we have all been in the state of paranga cetana, with our consciousness directed outwardly and thereby caught in the glamours of Maya, the glamours of the ever-becoming. Yoga, real yoga, is a slow and steady process of withdrawal.

The Maya will never go away. It is eternal. It, however, will slowly take on another meaning as its place in the eternal scheme of things slowly gets appreciated in a new way, and as one slowly, ever so slowly, tames their mind, and trains vairagya and viveka to see dispassionately the true order of things (Sanatana Dharma). Then, one is ready to sit and learn the asanas and breathing patterns that lead to samadhi, and a new stage in the adventure can begin.

So, in the meantime, just chill. Learn. Relax.

Appendix 1: Example of a Limit

Let's give a simple example for Readers who have not taken a calculus class. Say we have the following very simple equation:

$$y = \frac{1}{x} + 2$$

We want to know the value of y when x = zero and when x = infinity. Why? Because these are the two "ends" (like bookends) of the answer to the equation. When x = 0, x is at its smallest possible value and when x = infinity, it is at its largest possible value. In math-speak, this is called the "domain" of x.

We can substitute some values and see how it works out. We'll make a table with increasing values of x on the left, the value substituted into the equation in the middle, and the answer, y, in the right column. All the numbers are chosen so you can do all the math in your head, without a calculator.

x	1/x +2	y
0.001	(1/0.001) +2 = 1000 + 2	1002
0.01	(1/0.01) +2 = 100 + 2	102
0.1	(1/0.1) +2 = 10 + 2	12
1	(1/1) +2 = 1 + 2	3
10	(1/10) +2 = 0.1 + 2	2.1
100	(1/100) +2 = 0.01 + 2	2.01
1000	(1/1000) +2 = 0.001 + 2	2.001

As you can see, as *x* gets bigger, then 1/*x* gets smaller, and the answer to the equation approaches the number 2. Mathematically, it is then said that "as *x* goes to infinity, the limit of *y* = 1/*x* + 2 equals two". Mathematically this is written as:

$$\lim_{x \to \infty} \left(\frac{1}{x} + 2 \right) = 2$$

On the other hand, as *x* becomes a fraction less than one, then 1/*x* becomes a larger and larger number and two gets added to this increasing number. So, as *x* becomes a smaller and smaller fraction, from 1 tenth (1/10) to 1 one hundredth (1/100) to one one thousandth (1/000) and so on (the '...' I discuss in the article), then 1/*x* + 2 = 2 + increasingly big numbers. In short as *x* approaches zero, *y* approaches infinity. Then, the math is specified as:

$$\lim_{x \to 0} \left(\frac{1}{x} + 2 \right) = \infty$$

We can also make a graph of the answers to the equation:

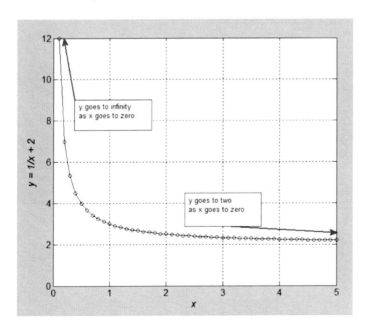

This shows as a picture what the equations describe in symbols. As *x* increases (follow the red arrow), the curve gets closer and closer to 2. As *x* decreases towards zero (follow the purple arrow), y increases forever up to infinity, so that at the point *x* = 0, the value of *y* = infinity.

This example makes it plain how you would have to solve an infinity of values of x into the equation to reach the limits. This is what I mean in the article that we cannot, in practice, solve an equation an infinity of times. This example also makes plain how intuitive the idea of a limit is. You can see very clearly that, as *x* keeps getting bigger, that 1/*x* will go to zero when *x* = infinity, and the limit of the equation will equal 2.

There is thus a weird kind of quantum jump that happens when taking a limit. We progress 1, 2, 3, ... but at some point just jump to infinity to get the final answer, and you can see how there is no other way for the answer to be. Which is to say, even if we can't calculate forever, we can intuitively just see what the answer should be. That is, at least for easy cases like this example. The general case is more complicated and one has to know much more math to solve the limit problems.

The basic idea expressed here was invented independently by both Leibniz and Newton in the mid 1600s. It's fair to say these Gentleman's ideas changed the world to what it is today.

Last point, if you have read this far. While this stuff seems like boring math stuff, it is exactly this kind of thinking that lets us build rockets, send spaceships to Mars, build stoves, refrigerators and microwave ovens, cars, assembly lines, robots and in fact, all the machines that characterize modern times from all other times in human history. This is the stuff scientists and engineers learn and the above is like "ga ga goo goo" (e.g. baby's first words) of the math and science language used to understand nature.

In short, these ideas are a total big deal.

Thanks for reading.

Other Books by Donald J. DeGracia, Ph.D.

Note that all my writings are available for free at: www.dondeg.com.

Print on demand versions are available for sale for people who like to hold and read real books (like me). You can also buy eBook versions for a nominal fee.

Beyond the Physical, A synthesis of science and occultism in light of fractals, Chaos and Quantum theory.

1994, 428 pages.
Buy it! ($20.06)
Get it Free!

The Western mind is enamored -nay - hypnotized with what it perceives outside of itself. But for all its knowledge of the outer world, the inner world of consciousness is but a hazy, half-felt realization in the life of the so-called "modern" person. Beyond the Physical seeks to go inside, and shed light on the relation between the inner and outer worlds of our human experience.

Do_OBE How to lucid dream, astral project and have out-of-body experiences

1997, 284 pages
Buy it! ($14.46)
Get it Free!

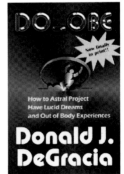

Finally in book form!!! Don DeGracia's massive internet hit: DO_OBE (pronounced do oh be eee). Millions of copies have been downloaded world wide. DO_OBE teaches how to LEAVE YOUR BODY!

Using straight forward, common sense language, DO_OBE is a guide to the inner realms of which our visible world is just the surface. Intelligent, fun and unpresumptuous, this book has been time tested and proven to be effective at teaching beginners the essential techniques to achieve the eluisive "out of body experience". SEE FOR YOURSELF FIRST HAND: Can you really leave your body? What can one expect to find in the inner realms of the mind? How do I do it??? It's all here!!!. This book is for all curious seekers who sense that there is more to reality than meets the eye, or the senses, or the world of our everyday life.

What is Science?

2014, 110 pages.
Buy it! ($39.99)
Get it Free!

In spite of the amazing technological marvels of the modern world that have stemmed from science, there is no agreed upon definition of what science is. In this lively, colorful, and engaging work, Don DeGracia contends that science is a very weak form of what has been described for thousands of years in Hindu India as "samadhi". Samadhi is an advanced technique of Raja Yoga in which the meditating subject fuses with the object of meditation, in a process that has been called 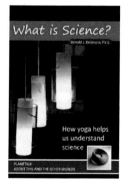 "knowing by being". By understanding science as a weak form of samadhi and comparing it to the knowledge aquired from yogic practices, many of the limitations of science are brought to the fore. These include: the link between mind and body, the role of the senses as middle-men between the mind and the objects of perception, why mathematics is "unreasonably effective" for describing the physical world, and how and why power is unlocked by the human mind when correct knowledge is obtained.

CPSIA information can be obtained
at www.ICGtesting.com
Printed in the USA
LVXC02n1728021214
416750LV00002B/29